PRACTICAL WASTE TREATMENT
AND DISPOSAL

PRACTICAL WASTE TREATMENT AND DISPOSAL

Edited by

DENIS DICKINSON

*Compiled in Collaboration with the
Institute for Industrial Research and Standards,
Dublin, Ireland*

A HALSTED PRESS BOOK

JOHN WILEY & SONS
New York — Toronto

PUBLISHED IN THE U.S.A. AND CANADA BY
HALSTED PRESS
A DIVISION OF JOHN WILEY & SONS, INC., NEW YORK

Library of Congress Cataloging in Publication Data

Dickinson, Denis.
  Practical waste treatment and disposal.

  "A Halsted Press book."
  1. Factory and trade waste. 2. Sewage—Purification.
I. Title.
TD897.D49   628.5'4   73–22498
ISBN 0–470–21278–0

WITH 14 TABLES AND 36 ILLUSTRATIONS

© APPLIED SCIENCE PUBLISHERS LTD 1974

Printed in Great Britain by Galliard (Printers) Ltd Great Yarmouth

LIST OF CONTRIBUTORS

S. W. BAIER, F.R.I.C., F.I.M., F.I.M.F.
Metal Users' Consultancy Service, The British Non-Ferrous Metals Research Association, London, England

R. W. BROOKS, B.Sc., C.Eng., M.I.C.E., F.I.Mun.E., M.Inst.W.P.C.
Assistant Engineer, Coventry BC, Warwickshire, England

M. C. DART, B.Sc., Ph.D., A.M.Inst.W.P.C.
Manager, Effluent Section, Engineering Division, Unilever Ltd., London, England

DENIS DICKINSON, M.Sc., Ph.D., M.Inst.W.P.C., F.R.S.H.
Consultant; Scientific Adviser to the Institute for Industrial Research and Standards, Dublin, Ireland

A. H. LITTLE, B.Sc., F.R.I.C., F.T.I., M.Inst.W.P.C.
Consultant on Industrial Effluent Treatments, Cheshire, England

E. H. NICOLL, B.Sc., F.I.C.E., F.I.P.H.E., M.Inst.W.P.C., M.I.W.E.
Assistant Chief Engineer, Scottish Development Department, Edinburgh, Scotland

T. WALDEMEYER, B.Sc., F.R.I.C., M.Inst.W.P.C., F.I.P.H.E.
Chemical Inspector, Directorate General of Water Engineering, Department of the Environment, London, England

A. B. WHEATLAND, B.Sc., A.M.Inst.W.P.C.
Head, Processes Division, Water Pollution Research Laboratory, Department of the Environment, Stevenage, England

S. WOLSTENHOLME, A.M.C.T., F.R.I.C.
Lecturer and Supervisor Experimental Tannery, Procter Department of Food and Leather Science, University of Leeds, England

v

CONTENTS

INTRODUCTION

When compiling a book on the treatment and disposal of wastes and waste waters from modern communities, one is soon faced with the impossibility of covering all industries under all kinds of circumstances. New industries are appearing continually, producing wastes and effluents containing new substances, often of unknown properties in the context of environmental pollution. Consequently, it becomes necessary to fall back on principles for the guidance of those faced with new problems. In the present volume we have devoted the first two chapters to these principles; the first outlines the evolution of a system of control of water pollution; the second goes more deeply into the chemistry both of pollution and of the treatment of effluents by biological means.

A classification of effluents into four types has been evolved and, in general:

Class (1) effluents, whose effects are physical, will be improved by physical means;

Class (2) effluents, whose effects derive from their organic nature, will be improved or purified by biological methods;

Class (3) effluents, of chemically polluting character and often toxic, will be treated by chemical reactions; and

Class (4) effluents, which are both organically polluting and toxic, will require a combination of chemical, biological, and physical methods for their purification.

Class (1) includes such effluents as cooling-waters from power stations and sand and gravel washings. Little further attention has been given to them as their treatment will depend in each case on the site selected for disposal; and the chapter on principles will elucidate the requirements. Also, the treatment itself is usually quite simple and self-evident. In Class (2) we have a wide variety of effluents, including sewage. The treatment of effluents of this type is exemplified in special chapters devoted to food manufacturing effluents of various kinds, to farm wastes, and to the treatment of sewage from isolated communities. The connection between this last subject and industrial development may not be self-evident. It arises in developing areas: mining or construction camps; development of the tourist industry by the establishment of caravan sites and of hotels; holiday villages, and the like, isolated from any town; even from the establishment of a prison.

The chief examples of Class (3) effluents arise from the metal finishing industries, to which a chapter is devoted. This will be useful to those concerned with other industries that deal with chemicals, as the principles which have been put into effective use in the metal plating industry must find analogous application elsewhere. In Class (4) the most notorious industry is tanning, whose problems are extensively explored in Chapter 10. Certain other industries—such as paper and board manufacture and textiles—do not fall exclusively into any one class because of the variation in the processes carried out in different establishments.

The treatment and disposal of solid wastes is considered in each industrial chapter and in addition an account is included of the conversion of sewage sludge (or more exactly, of an *industrial* sewage sludge) into utilisable gas. This is a subject on which comparatively little has been written, and as a process its potential has not been fully explored. It is apparent, however, that industries producing a digestible waste (those in Class (2)) might well find sludge digestion worth while, always provided they can operate on a sufficiently large scale.

There are references in some of the chapters dealing with individual industries to legal standards and administrative anti-pollution procedures adopted in the United Kingdom or elsewhere. In general, however, we have not given too much attention to the legal system of any country, as the standards of purification required and the apportionment of responsibilities and costs between one body and another must be very much a matter of politics and economics—which are outside the scope of this manual.

EDITOR'S NOTE ON UNITS

The various contributors to this book have used metric and Imperial units, some have used both, more or less indiscriminately. It was suggested that Standard International metric units should be substituted throughout and all the appropriate conversions made. Past experience with standardisation makes us think that such an exercise would not be worth while. Once we had *grains per gallon*; then *parts per* 100 000; then *parts per million*; then *mg per litre*: some day somebody will decide that as *mg per litre* is not independent of temperature we must change again, possibly to *mg per kil*. One can only conclude that Standard Units are Standards only for the time being and are liable to change at any time. This is perhaps as well because the set which is fashionable at the moment is certainly not ideal: the cubic metre, for example, is too big a unit for the measurement of low flows and it is 1000 times greater than the next lower unit, the litre; so that we can anticipate the introduction of an intermediate measure at some time.

Apart from this inconstancy, however, there is the consideration that standardisation on a set of units tends to 'date' (*i.e.* imply 'out of date') work which was done in a previous era. There is already far too much neglect of work done more than ten years ago and we would not wish to encourage this tendency when a table of conversion factors is all that is required to make earlier units comprehensible. Several such tables are therefore appended to enable those who must to convert the cubic yards, gallons, Btu's, etc. into SI units.

CONVERSION FACTORS

Imperial to Metric (SI) Units

Imperial unit	Factor ($\times$)	Metric unit	Factor ($\times$)
in	$2\cdot54 \times 10$	mm	$\times 10^{-3}$
ft	$3\cdot048 \times 10^{-1}$	m)	$\times 10^{-3}$ $\Bigg\}$ $\times 10^3$
yd	$0\cdot144 \times 10^{-1}$	m)	
mile	$1\cdot609 \times$	km	$\times 10^3$
in^2	$6\cdot452 \times 10^2$	mm^2	$\times 10^{-6}$
ft^2	$9\cdot29 \times 10^{-2}$	m^2)	$\times 10^{-4}$ $\Bigg\}$ $\times 10^6$
yd^2	$8\cdot361 \times 10^{-1}$	m^2)	
acre	$4\cdot047 \times 10^{-1}$	hectare	$\times 10^4$
in^3	$1\cdot639 \times 10^4$	mm^3	$\times 10^{-9}$
ft^3	$2\cdot832 \times 10^{-2}$	m^3)	$\times 10^3$ $\Bigg\}$ $\times 10^9$
yd^3	$7\cdot646 \times 10^{-1}$	m^3)	
gallon (gal.)	$4\cdot546$	litre	$\times 10^{-3}$
lb	$4\cdot536 \times 10^{-1}$	kg	$\times 10^{-3}$
ton	$1\cdot016$	tonne	$\times 10^3$
Btu	$1\cdot055$	kJ	
hp	$7\cdot457 \times 10^{-1}$	kW	

To convert Imperial into the corresponding metric units simply multiply by the factor shown in the factor column; *e.g.* to convert inches to millimetres, multiply by $2\cdot54 \times 10$; to convert into a larger metric unit multiply also by the power of 10 shown under the factor column on the right-hand side of the table; *e.g.* to convert inches to metres, multiply by $2\cdot54 \times 10$ and by 10^{-3}, *i.e.* $2\cdot54 \times 10^{-2}$.

EXAMPLES

Conversion of pounds of Biochemical Oxygen Demand (BOD) per cubic yard to kilogrammes of BOD per cubic metre.

Express as a fraction, *i.e.*

$$\frac{\text{lb BOD}}{\text{yd}^3}$$

Substitute the conversion factors in this fraction,

$$\frac{\text{lb to kg}}{\text{yd}^3 \text{ to m}^3} = \frac{4\cdot536 \times 10^{-1}}{7\cdot646 \times 10^{-1}}$$

Divide to obtain the new factor, 0·5932, by which to multiply lb BOD/yd^3 for conversion to kg BOD/m^3.

In a similar manner British Thermal Units per cubic foot is converted to kilojoules per cubic metre by multiplying by the factor

$$\frac{\text{Btu to kJ}}{\text{ft}^3 \text{ to m}^3} = \frac{1\cdot055}{2\cdot832 \times 10^{-2}} = 0\cdot3726 \times 10^2$$

What volume of water is represented by one inch of rainfall over a hectare of land?

in to mm: 2·54 × 10
hectare to mm^2: × 10^4 × 10^6 = 10^{10}
volume of 1 in per hectare   = 2·54 × 10^{11} mm^3
or as m^3, multiply by 10^{-9}, = 2·54 × 10^2 m^3
or in litres, × 10^3 = 2·54 × 10^5 l.

To convert this back into gallons, *divide* by the factor 4·546.

Conversion Factors: Analytical

	Grains per gal.	Parts per 100 000	Parts per million[a]	Degrees Clark	Grains per US gal.	French degree of hardness	German degree of hardness
Grains per gallon	1	1·43	14·3	1	0·829	1·43	0·8
Parts per 100 000	0·7	1	10	0·7	0·583	1	0·56
Parts per million[a]	0·07	0·1	1	0·07	0·058	0·1	0·056
Degrees Clark	1	1·43	14·3	1	0·829	1·43	0·8
Grains per US gallon	1·2	1·71	17·1	1·2	1	1·71	0·96
French degree of hardness	0·7	1	10	0·7	0·583	1	0·56
German degree of hardness	1·25	1·78	17·8	1·25	1·04	1·78	1

[a] Synonymous with *mg per litre* and *mg per kg* for all practical purposes.

Degrees of hardness are expressed in terms of $CaCO_3$ *except* German degrees, where the basis is CaO.

To convert a value expressed in one unit into another, find the existing unit in the left-hand column and multiply the value by the factor given under the appropriate column head.

TEMPERATURE, CHLORIDE CONTENT, DISSOLVED OXYGEN

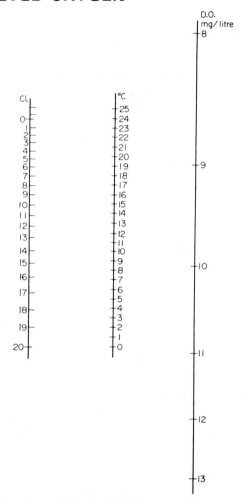

Nomogram relating temperature, chloride content (in g Cl/kg of water) and dissolved oxygen concentration (in mg/litre) in sea water/fresh water mixtures saturated with air. Based on data quoted by H. W. Harvey in *Recent Advances in the Chemistry and Biology of Sea Water*, Cambridge University Press, 1945.

A rule laid across the appropriate chloride concentration (left-hand scale) and temperature (middle scale) will intersect the dissolved oxygen scale (right-hand scale) at the corresponding figure.

Chapter 1

SCIENTIFIC PRINCIPLES OF THE CONTROL OF WATER POLLUTION*

DENIS DICKINSON

THE NATURE OF WATER POLLUTION

Chemically pure water does not exist in nature; water is too good a solvent. Freshly precipitated rain water dissolves gases from the atmosphere through which it passes, then it dissolves, more or less, minute proportions of everything on which it falls, or over or through which it flows. The result is that any river wherever it may be reflects in its chemical composition the history of the water of which it is made up.

Nature has provided series of interdependent organisms—animals and plants—which together comprise an ecosystem the overall function of which is to prevent the accumulation in waters of excessive concentrations of dissolved or colloidal substances. In this sense, all waters are polluted and it is often desirable that they should remain so. The lowest organisms in the ecosystem use the impurities as food, higher organisms feed on lower through the scale, so that in order to support life a certain amount of 'pollution' is essential.

As commonly understood, however, water pollution has a rather different meaning; it is thought of as the undesirable effect of too high a concentration of some noxious material. In fact, the difference between desirable and undesirable pollution is only a question of degree. Undesirable pollution arises as a result of an imbalance

* An explanation of the philosophy on which *Recommendations for the Disposal of Industrial Effluent Waters*, published by the Institute for Industrial Research and Standards, Dublin, is based.

in the ecosystem which normally thrives. A most important limiting factor in the natural control of water pollution by ecosystems is the quantity of dissolved oxygen available. Dissolved oxygen is essential for the aerobic oxidation or utilisation of pollutants by aerobic ecosystems. If the quantity of pollutant is greater than can be oxidised by the existing ecosystem and dissolved oxygen supply acting together, then the concentration of dissolved oxygen in the water becomes depleted and other ecosystems develop which utilise other sources of oxygen. The most common alternative source is sulphate which becomes reduced to sulphide, with unpleasant consequences. Before this stage is reached, however, the higher aquatic organisms —including fish—which require dissolved oxygen for their well-being disappear from the ecosystem.

The ecosystem which prevails at any time depends primarily on the level of dissolved oxygen concentration in the water and if it is required that this ecosystem shall include fish, a minimum acceptable concentration of dissolved oxygen is fixed automatically. This is generally agreed to be about 4 mg/litre in temperate climates. However, the more desirable species of fish—trout and salmon—require a higher concentration for their continued well-being (about 5 mg/litre) and a still higher concentration for breeding.

The type of polluting matter considered so far is that which encourages the growth of one sector of the ecosystem and produces, primarily, imbalance. The general case would apply even if that sector of the ecosystem which multiplied excessively were fish, as overpopulation with one or more species of fish, produced by an overabundance of food, would result in eventual pollution of the water.

The concept of excessive growth of a sector of the ecosystem introduces Time as a second limiting factor. Every biological entity considered in this context has a growth rate; it grows as it oxidises or transforms the polluting matter. Oxidation or transformation requires time for its completion. With very rare exceptions, natural waters flow in some direction and in any given time a quantity of fresh water will arrive at or pass any particular spot. This fresh water may contain none of the pollutant being added at that spot, but it will contain dissolved oxygen. If this quantity of dissolved oxygen is sufficient to allow the ecosystem to accomplish complete oxidation of the pollutant, then the dissolved oxygen content of the mass of water will fall only to a level determined by the rate of dissolution of fresh oxygen from the air and supplemented by aquatic plants. This explains why many discharges of polluting matter into rivers do not give rise to any of the undesirable effects associated with polluted waters. If a proper balance is preserved, the rate of depletion of dissolved oxygen by the oxidation and transformation of the pollutants is countered by the rate of replenishment of the dissolved oxygen. In the case where the rate of utilisation of the dissolved oxygen exceeds its rate of replenishment a zone of polluted water may appear after a time, but, in the absence of further additions of polluting matters, this will in turn disappear. This overall phenomenon is referred to as the capacity of the water for self-purification.

It is therefore quite possible to dispose of an unwanted material into a body of water without producing any ill-effects whatsoever, provided the nature of the pollutants is such that they can be oxidised and transformed by the prevailing ecosystems and their quantity is restricted to that which can be so oxidised or transformed at the same rate as the dissolved oxygen is restored. Broadly speaking—the quantity of pollutants added must not exceed the self-purification capacity of the body of water.

PRINCIPLES OF WATER POLLUTION CONTROL

The quantity of oxygen required to oxidise a pollutant is measured in the test known as the Biochemical (or Biological) Oxygen Demand (BOD). The ultimate oxygen demand can be measured by prolonging the test over ninety days, at the end of which time it is assumed that any matter which can be oxidised under the conditions of the test will have been so oxidised, but this is rarely done. It is usual to employ a period of five days at $20°C$ as this gives a result which is reasonably reproducible and can be used as a basis for calculations. In considering BOD values, however, we must never lose sight of the time factor involved, nor of another essential requirement of the test—that the dissolved oxygen content of the test water at the end of the five-day period should not be less than 4 to 5 mg/litre, *i.e.* approximately half what it was at the beginning of the test. When assessing the probable effect of a discharge of known BOD on a body of water, it is necessary to ensure that the dissolved oxygen content of the mixture of discharge and receiving water shall not fall below 4 mg/litre, not only to preserve fish life and avoid obvious pollution, but also to make one's calculations valid. Consequently, in order to arrive at a fairly certain assessment, it is desirable to know

(1) the dissolved oxygen concentration in the receiving water;
(2) the BOD of the receiving water;
(3) the quantity of fresh water arriving at the discharge point during an average day;
(4) the BOD of the polluting discharge;
(5) the daily quantity of the discharge;
(6) the duration of the discharge and its variation during each period of a day, and a week.

The dissolved oxygen content of the discharge itself is usually ignored, although in some instances it may provide a useful safety factor.

To take an example: a potentially polluting liquid of organic origin with a BOD of 500 mg/litre is to be discharged at a rate of 10 000 gallons per hour into a river which has a BOD of 2 mg/litre and a minimum flow of 24 million gallons per day. The dissolved oxygen concentration in the river water is 9 mg/litre. The dilution factor is

$$\frac{\text{The river flow per 24 h}}{\text{Effluent flow per 24 h}} = \frac{24 \times 10^{6}}{10^{4} \times 24} = 100$$

and the average BOD of the mixture would be

$$\frac{1 \times 500 + 100 \times 2}{1 + 100} = 7 \text{ mg/litre (approx.)}$$

It is apparent that the river below the discharge point, with the discharge thoroughly mixed in, would have a BOD of 7 mg/litre, so that the proposition is immediately dubious. The Institute for Industrial Research and Standards *Recommendations for the Disposal of Industrial Effluent Waters* require a maximum of 5 mg/litre for the BOD of such a mixture, stating that:

> The biochemical oxygen demand (BOD) of the effluent should be such that on admixture with the receiving water it will not increase the BOD value of the receiving water by more than one milligramme per litre. In no circumstances should the BOD value of the effluent be such that the BOD of the receiving water will be increased to more than 5 mg/litre.

Recommendations apart, would such a situation give rise to obvious pollution and/or loss of fish life? The hourly discharge would require 50 lb of dissolved oxygen to satisfy its five-day BOD. The dissolved oxygen content of the river water arriving at the point of discharge amounts to 90 lb *per hour* and its own BOD would consume 20 lb of dissolved oxygen. If, therefore, the whole of the mixture were to be impounded in some way and given no opportunity for further oxygen to dissolve from the air, 70 of its 90 lb of dissolved oxygen would be used up in 5 days, leaving a residue of 20 lb in a total volume of 1 010 000 gal (or in 10 100 000 lb) which is approximately 2 p.p.m., or 2 mg/litre. In this case it would be impossible for anaerobic conditions to develop in the river as a whole as a result of the hypothetical discharge and it is unlikely that serious pollution would arise.

Less favourable conditions might, however, prevail. The Recommendations require that the mixture of effluent and receiving water should have a maximum BOD of 5 mg/litre—less than in the hypothetical case stated—but also that the dissolved oxygen content of the river, considering a normal river, should be a minimum of 4 mg/litre. If, in this instance, the dissolved oxygen content of the river water fell for some reason from 9 to 4 mg/litre before it arrived at the point where it received the discharge of effluent, its contribution to the dissolved oxygen balance would then be only 40 lb per hour, which is less than the 50 lb required to meet the BOD of the effluent discharged. In this event, would the river become anaerobic?

Two further factors should be considered—time and the geography of the area. The five-day BOD will, by definition, require a period of five days to absorb the 50 lb of dissolved oxygen. Although the daily rate of absorption is not constant, it is unlikely that more than 20 lb would be absorbed in the first period of 24 h after mixing, unless the oxygen demand of the effluent were of a chemical rather than a biochemical nature. As to the oxygen demand of the river water itself, 4 lb is a likely figure for the same period. Consequently, of the 40 lb of dissolved oxygen conveyed by the river water, 24 lb may well be consumed in the first 24 h after mixing in the effluent, leaving a residue of 16 lb, or 1·6 mg/litre, in the mixture. Although undesirably low, this is still far from anaerobic and, moreover, presupposes that no further oxygen has dissolved from the air during 24 h. The forward velocities of rivers vary tremendously, but one mile per hour is low indeed. Assuming this to be the velocity, the mixture will have travelled 24 miles downstream from the effluent discharge point during this first day. It is extremely likely that over such a distance several minor, and possibly major, tributaries will have added significantly to the dissolved oxygen content of the river; and this is in addition to the oxygen which must dissolve from the air through the surface of the water. This latter process is greatly accelerated by turbulence, however caused. Without calculating the probable surface area of a 24-mile stretch of river and making various assumptions to calculate the probable weight of oxygen dissolved *via* this surface, it must be obvious that in the case stated there is no real danger of the river becoming either anaerobic or unsuitable for fish. And in arriving at this conclusion we purposely ignored two of the Recommendations.

It follows that the strict application of the Recommendations of the Institute for Industrial Research and Standards *must* (a) prevent pollution where none exists, and (b) improve the river water quality when it is already polluted. In order to state the case more clearly, the argument has been allowed to develop from the general situation—which covers all bodies of water usable as receptors—to the particular case of a river. This does not imply any change in principle: the fundamentals of the Recommendations remain unchanged.

CLASSES OF POLLUTING EFFLUENTS

The Recommendations group effluents into four classes, as follows:

Class (1) Non-toxic and not directly polluting but liable to disturb the physical nature of the receiving water.

Class (2) Non-toxic, but polluting by reason of an organic content of high oxygen demand.

Class (3) Toxic; effluents containing directly poisonous materials.

Class (4) Polluting by reason of organic matter of high oxygen demand and toxic in addition.

So far the type of pollution envisaged has been that arising from Class (2) effluents, *i.e.* those which contain solids in suspension which can be readily consumed by fish, insects, or birds, and/or organic matter in solution which is readily utilised by micro-organisms. Domestic sewage and wastes from food factories and creameries are the main source of effluents of this Class. Class (3) effluents—'Toxic; effluents containing directly poisonous materials'—include sewage from industrial areas as well as effluents from individual factories carrying out chemical processes. These contain substances which will deter or kill fish, animals, and birds, will prevent microbiological action and, in the worst cases, effectively sterilise a body of water. They unbalance the ecosystems in a negative way by eliminating one or more sections. Thus they may have a secondary effect by permitting the proliferation of less desirable sectors of the ecosystem and will certainly interfere with the self-purification capacity of the water body.

The aim of the control of such effluents must be to ensure that the concentration of their toxic constituents in the mixture of effluent and receiving water never reaches the toxic threshold. At first sight this would appear to be more simple to calculate than the effects of Class (2) effluents although the arithmetic is obviously complicated by the same considerations of relative and minimum flows. However, it is facilitated in that we do not have to take the time factor into account. If we can ensure that the concentration of toxic elements in the mixture of industrial effluent and receiving water is acceptable immediately after mixing, we can be assured that it will not increase at any point downstream (unless there is an additional discharge). We are also justified in assuming that the toxic concentration will *decrease* progressively as the mixture moves downstream, mainly because of dilution by other water, but additionally because of chemical reactions which serve to oxidise or complex most of these substances into harmless derivatives.

TOXIC SUBSTANCES

The decision as to the quantity of toxic substance which may be discharged into a body of water at any point must be based on a knowledge of the toxicity of the substance concerned, together with a full knowledge of all the other substances already present in the receiving water and their toxicities. The Recommendations go into this matter fairly thoroughly in so far as toxic metals are concerned. The limits, or allowable concentrations, of the metals in potential water supplies are:

Lead	not more than 0·05 mg/litre
Chromium	not more than 0·05 mg/litre
Cadmium	not more than 0·01 mg/litre
Copper	not more than 1·0 mg/litre
Zinc	not more than 5·0 mg/litre

With the exception of lead the various metals are more toxic or harmful to fish than to human beings. If the water is suitable as a fishery as regards its content of metals except lead it will in general be suitable as a water supply.

The toxicity of a metal to fish depends on the temperature of the water and also on the hardness of the water. If more than one toxic metal is present, the tolerable concentration of each metal is less.

The contents of metals in the fishery water should not exceed the values stated below for water having a hardness as indicated.

Water of less than 40 mg/litre hardness:

Lead	0·5 mg/litre
Copper	0·04 mg/litre
Zinc	0·4 mg/litre
Cadmium	0·01 mg/litre
Nickel	5 mg/litre
Chromium	1 mg/litre

Water of not less than 40 mg/litre hardness:

Lead	1·0 mg/litre
Copper	0·1 mg/litre
Zinc	1 mg/litre
Cadmium	0·01 mg/litre
Nickel	5 mg/litre
Chromium	1 mg/litre

In the event of more than one metal being present, a guide to the toxicity to fish of the water as a whole may be obtained by adding the ratios:

$$\frac{\text{Actual concentration of metal}}{\text{Limiting concentration of metal}}$$

the sum of which should be less than one if the water is to be considered safe.

Toxicity, however, is like pollution in general, a relative term. Many people would like to be given a *figure*, a *standard* which may not be exceeded; but it is just not possible to say that 'x mg/litre will kill all the fish, but $x - 1$ mg/litre will be safe'. Nor is it possible to take into account every eventuality, nor has the toxicity of every possible substance been measured. The most that can be done in practice is to list the acceptable concentrations of the most usual toxic substances, to calculate how much of a particular material might be added to a body of water *without* its becoming definitely poisonous, and then arrive at a final—and preferably lower—figure by taking into account geographical and social factors.*

There are exceptions, however, even to this general rule. For example, certain rivers do not support fish life and cannot be used for domestic purposes because they already contain high concentrations of toxic metals derived from the minerals in the area through which they flow. It would be pointless in such a case to prohibit a discharge into a river on the grounds that the effluent contained toxic metals; but there might be grounds for prohibiting a discharge of domestic sewage into the same river because the water may have a very small self-purification capacity and the sewage would cause fouling of the watercourse.

Exceptional cases of this nature are taken care of by adopting the principle that due regard shall be paid to the purposes for which a receiving water is, or is likely, to be used. This avoids the anomaly of protecting non-existent fisheries or totally unsuitable water supplies. Geological, geographical, and social factors are predominant in this area.

DISCHARGES TO ESTUARIES
The most difficult problems are posed by proposals to make discharges into tidal estuaries. In very few cases is sufficient known of the flows of fresh and sea waters in the estuary to permit a proper solution of the problem. It is not generally realised that the sea water entering an estuary as the tide rises mixes to only a limited extent with the fresh water supplied by the river itself. Because of the difference in density, fresh water tends to float on top of sea water, with the result that the salinity of the water in an estuary varies with the distance along the line of the estuary and with the depth. In many cases it will also vary across the estuary. In the absence of sufficient information to map the tidal wedge it is probably safest to assume that there is no mixing at all of the fresh and sea waters and that the tide merely acts as a barrier to the flow of the river out to sea. The sea can be visualised as a dam which moves up and down the

* The matter of toxicity and its assessment is dealt with in greater detail in Chapter 2. See also the note added in proof on p. 12.

estuary with the tides. Some of the fresh water will spill over this hypothetical dam, the maximum spill taking place at high tide, and this spill-over will be carried out to sea as the tide recedes. At low tide, the estuary may be occupied completely by fresh water from the river and this is effectively pushed backwards up the estuary as the tide rises. This accounts for the widely reported phenomenon of polluting matter being pushed back by the tide.

To arrive at a probable picture it is necessary to know the flow of fresh water in the river, the area of the estuary, and the tidal rise. If the estuary is regarded as the classical bath being filled by one tap and emptied by another we have an over-simplified, but workable, analogy. The tap filling the bath is the river; the drain-tap is the sea outlet. When the tide is coming in, the drain-tap is closed; when the tide is going out, the drain-tap is open. To what extent the 'bath' is emptied by the receding tide depends on the relative flows of the river, through the sea outlet, and the capacity of the estuary itself. The only item which is approximately constant in the actual situation is the area of the estuary; its capacity is a function of the height of the tide, which is usually variable, but often predictable from local tide-tables. Although in general the volume of water contained in an estuary is so large as to offer a very big dilution factor to any proposed discharge of polluting matter, it is practically impossible to be certain how much of this volume will be effective either as a diluent or as a medium for transport of the effluent out to sea. However, if the calculations are based on the flow of fresh water only, a safe conclusion can be reached.

A further complicating factor is the effect of sea water on the BOD. It is common knowledge that sea water has a certain preservative or disinfectant effect because it suppresses the activities of many of the saprophytic organisms present in fresh water. Consequently, the five-day BOD of an organic waste mixed with sea water may be less than that of the same waste mixed with fresh water. This is obviously not due to any change in the oxidisable nature of the waste, but results from retardation of the *rate* of oxidation. This means that the organic matter will take longer to oxidise; it also means that the daily demand on the dissolved oxygen content of the water will be lower and therefore the actual concentration of dissolved oxygen will be higher than would obtain in a corresponding mixture of the waste with fresh water. Whether these effects constitute an advantage or a disadvantage obviously depends on local conditions and particularly on the probability of any undecomposed waste being returned either into the estuary itself or on to neighbouring coasts. Much depends also on the nature of the waste; particles of vegetable matter such as are discarded by a food processing factory provide food for certain fish and birds and if they can be conveyed in a fresh state by sea currents into the areas where the animals can consume them, it is unlikely that any pollution will arise. If, however, they are deposited in quantity, more or less unchanged, on to a neighbouring shore, they are likely to create a nuisance either directly, or indirectly through the breeding of insects, or even by attracting excessive numbers of birds.

INFLUENCE OF CLIMATIC CONDITIONS

Although the above arguments have been developed primarily on the basis of experience gained in temperate climates, they are still valid in other zones provided that the differences due to climate are taken into account. The solubility of oxygen in water, for example, decreases with rising temperature, while the rate of utilisation of oxygen by microbial species increases with temperature. In warmer water, therefore, the effects of a polluting discharge may be expected to become apparent more quickly than in cold water. This applies equally to discharges into rivers, lakes, or the sea. One compensating factor appears to be that tropical fish do not require such high levels of dissolved oxygen concentration for their well-being as cold-water fish, although if the concentration is expressed on a 'percentage saturation' basis, there is probably little difference between the requirements of the fish. The factors apply in reverse in cold climates —there is more dissolved oxygen available but its rate of utilisation by micro-organisms is lower. Consequently, the effects of a polluting discharge may not appear for a longer time. In extreme cases pollutants have been known to be preserved for very long periods by becoming frozen in rivers and lakes, and their effects have not then shown until the water thawed and the temperature rose sufficiently to encourage bacterial growth.

To help assess effects under unfamiliar conditions it might help to modify the BOD test by conducting it at the temperature prevailing in the water under consideration and for longer or shorter periods, as applicable, than the standard five days. This is reverting to the original purpose of the test which was to assess the effect of a discharge on the oxygen content of a receiving water, using the actual water as diluent in the test—not forgetting a suitable 'blank' determination, of course.

This short review has pointed out some of the major difficulties involved in deciding on the probable effects of a discharge of waste matter into water at a particular point. In every case it is fair to say that there is a calculated risk involved. In the past this has not always been appreciated and in very many cases a degree of success has been apparent which has subsequently proved to be illusory. This is because the effects of water pollution may be separated from their cause by very considerable distances in both space and time. The associated phenomena of secondary pollution and eutrophication have been clearly recognised only in recent years. There are doubtless other important phenomena still awaiting discovery. Realising our limitations, a cautious approach is essential and a slavish addiction to standards or limits should be avoided; and in spite of caution some mistakes will be made. In order not to impede progress unnecessarily it is essential that decisions shall be based on the maximum of information subjected to a skilled critical examination.

NOTE ADDED IN PROOF

Many countries are co-operating in the maintenance of an International Index of Toxic Materials. Current information should be sought from national organisations such as the Institute for Industrial Research and Standards, Dublin, Ireland, and the Water Pollution Research Laboratory, Stevenage, Herts., England.

Chapter 2

CHEMISTRY IN WASTE TREATMENT AND DISPOSAL

DENIS DICKINSON

WASTE DISPOSAL IN NATURE

In a Natural Society the waste materials produced by one organism are used as raw material by some other organism in such a way that the accumulation of dead organic matter is avoided. Numerous species of organisms are involved in a perpetual series of alternate degradation and synthesis; they are mutually dependent on each other and a balanced biological society becomes established. Only if one species produces a non-utilisable waste product does the complex cycle become temporarily halted—temporarily until such time as the environment is invaded by new species capable of utilising the intractable material.

The interventions of Man have caused changes in environments and interruptions of the classic carbon and nitrogen cycles; the more advanced the community, the greater the interruption of the natural processes of waste utilisation. All civilisations have at some time reached a point where communities have been obliged to make provision for waste disposal and they have organised systems in imitation of natural processes—most usually composting and farming. The industrial developments of the present century have forced advanced communities to take an ever-increasing interest in the whole subject of waste treatment and disposal with a view to accelerating the natural processes already adapted. Thus we have seen the development of the modern technology of effluent treatment and disposal based on accelerated biological oxidation of materials with the general aim of producing reasonably pure effluents and innocuous solids.

In spite of the fact that the processes used are mainly biological, the development of the science has been due mainly to the activities of chemists. It is true that engineers have contributed very largely to the design, construction, and operation of plant, and that biologists have more recently begun to study the subject seriously, but most of the fundamental studies have been carried out by chemists. This has been for the very good reason that it was the analytical chemist who was the only professional man able to measure the effects of the biological processes involved and to assess the improvement, or otherwise, introduced by modifications to a process or plant.

BIO-FILTRATION

The two biological processes most widely used for sewage purification are biological filtration and the activated-sludge process. A biological filter—the earliest form of plant used widely for the purification of town sewage—consists essentially of a pile of rocks, stones, coke, slates, flints, or similar material (the medium) over which the polluted liquid is allowed to trickle. Air passes up through the medium as the waste trickles down. Organisms of very many kinds establish themselves in the filter; some—bacterial growths, moulds, yeasts and mosses—affix themselves more or less firmly to the medium; others—worms, insect larvae and hoppers—migrate over the growths, feeding as they go. Still others—flies and spiders— are there only incidentally, attracted to the filter as a breeding ground and by the food which it offers. A biological filter is thus a large specialised ecosystem supported on a firm foundation and existing more or less in isolation. Its population is variable according to the chemical composition of the wastes passing through it, and also on the prevailing temperature and climate, and on the rate of application of the waste.

More recently, plastics media for filters have been introduced. These have a number of physical advantages, of which light weight is one of the most important, but they also offer a bigger ratio of free space to volume than can be assured by the more solid traditional filter media and consequently they can be 'dosed' at a much higher rate; that is to say, the rate of application of waste water to a plastics filtration unit is much greater than is possible with a traditional filter. Here arises a fundamental difference between the two kinds. The ecosystem which becomes established on a stone filter includes many forms of life, each feeding on another, and the successful operation of the filter relies in no small measure on insects, worms, and hoppers to control excessive bacterial growth which would otherwise cause blockage—interfering with both the downward passage of the liquid and the upward flow of air. The greater flow rate applied to a plastics filter unit discourages insects and is more favourable to bacterial growth. The established ecosystem is therefore different and the greater *quantity* of biologically active growth (slime) is accommodated in the larger volume of free space which the design of the plastics medium provides. Excess growth—that which would otherwise fill the free space and prevent air circulation—is removed by the physical effect of the greater flow rate. Maintenance of a minimum flow rate and its efficient distribution therefore become vital factors in preventing blockage and the consequent development of anaerobic conditions. One consequence of the different biological systems prevailing in the two types of filters—solid media and plastics media—and the different mechanisms for the removal of excess growth is the greater quantity of unstable secondary sludge which is voided by the plastics units. Provision must be made for the prompt removal and disposal of this material.

ACTIVATED-SLUDGE PROCESS

In many ways the high-rate filter is more nearly akin to the activated-sludge process than to a biological filter. If an aqueous waste of organic origin—town sewage, dairy waste, dilute molasses, for example—is aerated, bacteria will grow in it. The dissolved organic substances are gradually converted into bacterial cells, carbon dioxide, and ammonia. Settlement after a period will concentrate the organisms in the lower layers of the vessel and the 'spent' supernatant liquid is then removed and replaced by fresh nutrient. If this process is repeated at intervals a mixed culture of organisms is built up. When the liquid under treatment is sewage, no further inoculation is necessary, but industrial wastes do not always contain a sufficient variety of organisms to give the desired mixed population and inoculation with, for example, fertile soil may well be necessary and certainly advantageous. Eventually an 'activated sludge' is obtained which consists largely of bacterial masses (zoogloea) and sustains many species of flagellates and protozoa. Essentially, the bacteria grow on the dissolved impurities in the waste water under aerobic conditions, and the higher organisms consume the bacteria or one another, so that overall there is transformation of colloidal and dissolved impurities into discrete cellular material and eventually into simple inorganic compounds: CO_2, H_2O, NH_3, HNO_3, etc.

As already stated, activated sludge is 'built up' or developed over a period of time. It is a material which is changing continuously with the conditions prevailing. When the waste fed to it is of constant composition; when the reaction time is constant; when there is always an excess supply of air; and the temperature is constant, the activated sludge will assume an equilibrium composition in regard to balance of species and inorganic concentration. In practice, this constancy of conditions is rarely, if ever, achieved and activated sludge is consequently variable in composition even within the same plant treating the same waste product. It has been suggested that activated sludge in a sewage treatment plant is also subject to seasonal fluctuations in composition, but this is almost impossible to prove. Having regard to all the variables involved, it is surprising that this most useful material does behave in a predictable manner in some respects.

THEORY OF ACTIVATED-SLUDGE PROCESS

For many years it was believed that purification by the activated-sludge process proceeded in distinct steps—adsorption, oxidation of carbonaceous materials, and oxidation of ammonia and its derivatives to nitrate. Adsorption is certainly the first major action and will be referred to again later. Oxidation of carbonaceous material begins almost immediately after mixing of activated sludge with its substrate and can be followed in laboratory and plant experiments by dissolved oxygen, respirometer, or E_h measurements. Removal of ammonia, however, seldom occurs to any significant extent in an activated-sludge plant. Only after prolonged aeration—periods of three days or more—as in the oxidation ditches constructed in Holland, does complete removal of ammonia occur. Added to these observations, we have the well known phenomenon of the BOD/ Time curve which traces two distinct steps or phases of biochemical oxidation, the second of which corresponds with nitrate production. The evidence was therefore in favour of a theory that oxidation of ammonia was a separate stage in the process and it has been argued that this stage was in some way inhibited by organic matter, which must therefore be removed first. A re-investigation of the matter at the Water Pollution Research Laboratories, Stevenage, revealed that the nitrification process was susceptible to classical physico-chemical treatment. It takes place in two stages. *Nitrosomonas* convert ammonia into nitrite, and *Nitrobacter* oxidise this further to nitrate. The rate of the whole process is controlled by the first stage, which is the slower. The growth rate of *Nitrosomonas* at 15°C results in an 18% increase in numbers per day. An equation from which the degree of nitrification obtainable under working conditions can be estimated was published by the Laboratory in 1963. This is:

$$t = \frac{24L(0\cdot1 + 0\cdot9t^{0\cdot5})}{(1 + p)S(0\cdot18)\,e^{0\cdot12(T-15)}}$$

in which t = the minimum period of aeration, in hours, for complete oxidation of ammonia to nitrate.

L = the 5-day BOD of the waste being fed to the plant.

p = ratio of sewage flow to return sludge flow.

S = concentration of sludge solids in the mixture (suspended solids).

T = temperature, °C, of the mixture.

$0\cdot18$ = estimated growth constant of *Nitrosomonas*, in reciprocal days.

It is assumed that a sufficient flow of air, the source of the oxygen necessary, is available at all times.

The solution of this equation is laborious, but not difficult if graphical methods are used to smooth out approximations. Alternatively, a computer can do the job more rapidly. The utility of the equation is best illustrated by considering its application to a particular plant.

CONTROL OF AN ACTIVATED-SLUDGE PROCESS

In the operation of sewage purification plant using the activated-sludge process, there are few variables which can be controlled. Effectively, t in the equation is the residence time in the plant, which is a reciprocal function of the rate of flow of sewage into the plant. If this is less than the value required for complete nitrification as indicated by the equation, then nitrification cannot be expected. L, the 5-day BOD of the influent is varying continuously, but the mean value over 24 hours during dry weather is reasonably constant and ascertainable. In the plant under consideration, the mean value of L was 180 mg/litre. The value of p could be varied to some extent, but an increase in p caused a decrease in t unless the flow of returned sludge was decreased to compensate; the extent to which this could be done was limited by consideration of final settling tank performance and the maintenance of sufficient sludge in the system. A value of $p = 1$ was used for most of the time. S, the suspended solids concentration in the mixed liquor, could be controlled by wasting the appropriate daily volume of excess sludge. T is a natural variable and no large plant yet constructed has means for temperature control. In order to ensure complete nitrification in a plant it is necessary to know what concentration of solids (activated sludge) is required at the temperature prevailing to allow the reaction to be completed during the time available—which is related to the flow rate. By constructing a nomogram from the equation, T, S, and t can be related as shown in Fig. 2.1. The residence time is determined by the

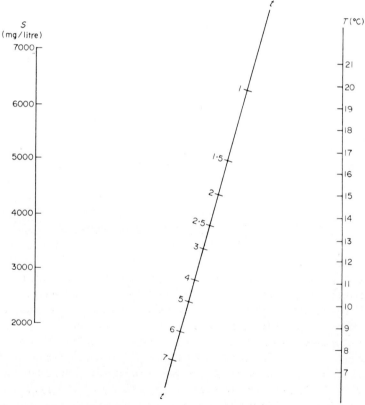

FIG. 2.1   Nomogram for control of nitrifying activated-sludge process, based on WPRL Equation in which $L = 180$ and $p = 1$.

flow rate, by p, and by the volume of the aeration plant. By laying a rule across from the temperature on the right-hand vertical scale to the concentration of suspended solids on the left-hand vertical scale, the intercept on the time scale gives the minimum number of hours required for completion of nitrification. Provided that this is less than the residence time, complete purification and nitrification should be attained.

After a few weeks of experiment, it became obvious that the predicted times were too low as nitrification was not achieved when theory predicted that it should be. Two possibilities existed; either the growth rate constant, 0·18, was too high for the prevailing conditions (WPRL experiments showed this to vary from one environment to another), or the assumption that S was a satisfactory measure of the concentration of *Nitrosomonas* was unjustified. Because the activated sludge in this particular plant contained a high concentration of inorganic matter (25% ash), the latter possibility was preferred—but this decision was also influenced by the practical impossibility of determining an alternative growth-rate for *Nitrosomonas*. In any event, multiplication of the predicted time by 2 was found to give close agreement between theory and practice.

Although temperature in a municipal plant in England varies with the season of the year, it does not change suddenly; there is an annual cycle, generally from a minimum of about 7°C to a maximum around 19°C. Under conditions of dry-weather flow a mean maximum residence time can be calculated. Suppose this to be 3 h. Incorporating the S factor of 2, this corresponds to 1·5 h on the nomogram. At a temperature of 19°C, any solids concentration greater than 3300 mg/litre should give complete nitrification. As the temperature falls complete nitrification can still be achieved if the suspended solids concentration is allowed to increase. There is, however, a practical maximum concentration for most plants at around 7000 mg/litre and this would cause loss of nitrification to occur when the temperature fell below 13°C.

Experiments with actual plants are necessarily slow as one is not free to make changes just to prove a point. However, over a period of about a year sufficient opportunities occurred to cover a wide range of variables and confirm the general applicability of the equation—modified by the substitution of $\frac{1}{2}S$ in the denominator instead of S, to suit the characteristics of the plant in question. As a result we are satisfied that concurrent oxidation of carbonaceous

matter and of ammonia are readily obtained when the concentration of the requisite organisms—as reflected in the sludge concentration—is sufficiently high. There are consequently no grounds for
maintaining that the carbonaceous matter is in some way inhibitory
to the growth of *Nitrosomonas* and consequent nitrification. Division
of the purification process into more than two consecutive stages is
not warranted. The failure of older activated-sludge plants to
produce nitrified effluents from sewage is due mainly to their being
run at too low a solids content—although it would not be true to
say that an increase of activated-sludge concentration to a level of
5000 mg/litre or above is the only step necessary to improve their
performance. It is a condition of success that at no time must the
reaction velocity be limited by the oxygen supply. However, it *is*
true that if the effluent from an activated-sludge plant contains
appreciable concentrations of nitrate—4 mg/litre or more—a very
high degree of purification is being obtained and the residual BOD
of the effluent is likely to be much less than 20 mg/litre. This is not
because there is necessarily any standard sequence to the reactions,
but because the concentration of the reactant—activated sludge—
which is necessary to bring about nitrification automatically ensures
a high concentration of those organisms which oxidise carbonaceous
materials. This general rule has also been found to apply to
laboratory-scale experiments on the treatment of dairy and slaughterhouse wastes, and there is no reason to doubt its applicability to
large-scale plants, always provided that there is at all times an
excess supply of air. In warmer climates correspondingly lower
concentrations of sludge solids should be effective, whilst in colder
climates the converse would be true.

STAGES IN ACTIVATED-SLUDGE PROCESS
The biophysical phenomenon of the first stage of the activated-
sludge process, adsorption, may now be considered further. It was
realised many years ago that activated sludge is an effective adsorbent
for organic substances and there are processes designed to make good
use of this property. Perhaps because most of the research work was
done by chemists, some of whom tried to make activated sludge
conform to physico-chemical laws, contradictory claims were made
from time to time. These can, however, be reconciled when it is
accepted that the properties of a sludge are a reflection of the
biological activities of the organisms it contains, and that these in

turn depend on the waste which it is treating and the conditions—
particularly aeration intensity and temperature—prevailing. The
present author has summarised the activated-sludge process in terms
of redox potential as follows:

$$E_1 \;—\; \text{Adsorption} \;\rightarrow\; E_2$$

$$\begin{array}{ccc}
\uparrow & & | \\
\text{Reactivation} & & \text{Oxidation} \\
| & \text{Desorption,} & \downarrow \\
E_4 & \leftarrow \;\text{Utilisation,} \;—\; & E_3 \\
& \text{etc.} &
\end{array}$$

In most plants separation of the activated sludge from the purified
liquid took place at stage E_4 and the reactivation step was generally
minimal, in terms of time. It was later realised that separation of the
sludge at stage E_2 offered advantages. This developed into 'partial
treatment activated sludge'. In a plant operating this system a very
rudimentary sludge built up, consisting almost entirely of bacterial
flocs and usually devoid of protozoa. The physical properties of this
material are not suitable for a complete treatment plant, but it
works extremely well as an absorbent, removing the more easily
oxidised materials and thus reducing the load imposed on the
second stage of purification, which is carried out in bio-filters or a
secondary activated-sludge plant. The build-up of this rudimentary
material is relatively large so that there is a large surplus for separation at stage E_2 for further treatment by anaerobic digestion.
Adoption of this system saves considerably in plant capacity and
aeration time. It is a system well-suited for the treatment of industrial
sewages which contain heavy metals, cyanides, phenols, and other
industrial chemicals which may interfere with oxidation processes.

If activated sludge is regarded as a mass of enzymes and the
mixture of sludge and substrate as constituting a single (though
complex) redox system, the above cycle predicts that

$$E_1 > E_2 = E_3 < E_4$$

Plotting E against time of aeration we should obtain a curve as
shown in Fig. 2.2. A single-stage purification plant has to be large
enough to contain and aerate the mixture of waste and activated
sludge for sufficient time for completion of the cycle, i.e. t_4. If the
sludge were to be separated after time t_2 and thereafter aerated
separately for the period t_4–t_2, the plant would need to contain

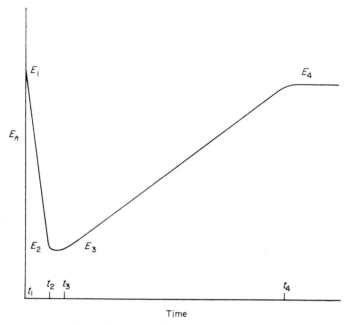

E_h

Time

FIG. 2.2   Redox/time curve (theoretical) for purification process.

only the sludge for this period and the volume of the plant required would be very much smaller than for a single-stage unit.

CHANGES IN REDOX POTENTIAL

Measurements of oxidation-reduction potential in a sludge/substrate mixture are not easy either to reproduce or to interpret. *Changes* of redox potential with aeration time, however, give valuable information. An account of laboratory experiments on this subject has been published by the author and may be summarised as follows. First, it is considered that the volume of sludge used should be quite large—not less than about 20 litres; aeration must be vigorous and stirring efficient. The difference in potential between a bright platinum electrode and calomel half-cell is measured, amplified, and recorded on a chart. For best results the concentration of sludge must be in the region of 6000 to 7000 mg/litre and it must be in a nitrifying condition. Given these conditions, substrates which the sludge could readily assimilate produced traces of the shape expected. An excellent example was a dilute waste from a meat products factory (Fig. 2.3); successive and increasing volumes of this gave curves which enclosed increasing areas, from which a relationship between the chart area enclosed by the curve and the work done by the sludge in purifying the waste was apparent. A plot of this area against the volume or weight of material oxidised gives a line or curve which is characteristic of the waste and of the sludge *at that time*. Unless measurements are made in quick succession and with comparatively small quantities of the waste material, the relationship

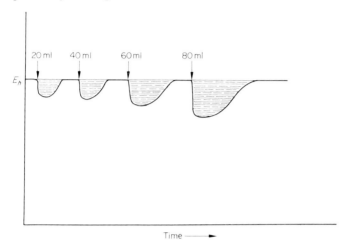

FIG. 2.3   Redox/time curves—increasing successive quantities of meat factory effluent added to laboratory activated-sludge tank.

between chart area and quantity of material is bound to change as a direct result of changes in the biological composition of the sludge induced by its own activity in purifying the materials in the waste. This results in curvature of the line. If it curves towards the A (area) axis, this indicates that the substrate is affecting the activity of the sludge adversely. Given time, the composition of the sludge may change to compensate for this, which is the same as saying that the sludge will adapt to the substrate: but if it does not do so, the material can be added only in limited quantities, preferably in admixture with some more suitable substrate.

AUTOMATED APPARATUS

It often happens that the redox curve does not quite return to its original base-line, *i.e.* there is a permanent shift in the E_h of the

activated-sludge system. This effect can be attributed to the accumulation of end-products, particularly of nitrate, or to a shift in pH
value. It was at one time thought that the redox/time curve area
resulting from the addition of a standard quantity of a reference
chemical such as glucose might be used as a measure of the activity
of the sludge system, but it has not so far proved to be reliable. Much
greater success was achieved by automating the whole system and
following the changes in the shape and position of the traces during
prolonged series of additions. The mechanics of this are simply to use
a rotary cam timer programmed as follows:

Time:	0	1	3	48	58	60 min
Aerator:	On ——————— Off					
Pump 1 (addition)		On—Off				
Pump 2 (abstraction)				On——Off		

During each hour a volume of substrate was added to the aerated
mixture by a peristaltic pump operating for two minutes. Forty-
eight minutes after the first addition the air was switched off and
after 10 minutes' settlement a volume of purified solution equal to
the volume of substrate added was removed from the system by a

Fig. 2.4   Laboratory-scale activated-sludge process: automatic addition and
extraction with recording of the redox, for assessment of the treatability of trade
effluents. (Apparatus developed at the Institute for Industrial Research and
Standards, Dublin.)

second peristaltic pump operating during the 59th and 60th minutes (Fig. 2.4).

The results of automation were far better than might be anticipated, particularly as it was found that the non-intervention of manual operations prevented certain 'static' effects. It became possible to run the process continuously for long periods, to relate the weight of activated sludge produced to the quantity of waste purified and to measure the efficiency of oxygen utilisation. It was found that previous ideas of the time required to purify organic wastes were often erroneous and that the rate of purification was generally much higher than expected.

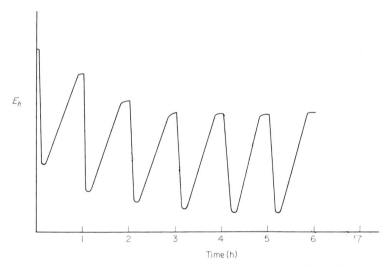

FIG. 2.5   Change in redox/time curve: automatic dosage, 1-h cycle, waste slightly too strong initially.

Irrespective of what it actually measures, the redox trace gives a clear indication of the start and end of the reaction and eliminates the need for time-consuming chemical control tests. There are two approaches to measuring the treatability of a waste of unknown properties. One is to arrange for serial additions of a fixed volume containing rather more material than the sludge is able to oxidise within the time cycle and to allow the concentration of sludge to increase to a level which *can* complete the purification. In this event the redox traces are expected to change as shown in Fig. 2.5. The

other method is to adjust the volume or concentration of the waste
fed so as to achieve complete purification within the time cycle.
Sometimes a combination of both methods has to be used. The
effect of an additive—detergent, metal salt, etc.—can be assessed by
tracing the effect of its addition in known concentrations to a standard
nutrient. For this purpose dried whey or skim-milk powder made up
to a concentration of about 1% in water and fortified with 5 or
10 mg/litre of ammonia has been found useful. In general terms 1 g
of a suitable nitrifying activated sludge (dry weight) purifies about
0·02 g of skim-milk powder in one hour.

SIMILARITY OF HIGH-RATE FILTRATION AND THE ACTIVATED-SLUDGE PROCESS

If we now consider high-rate plastics filter units in the light of the
redox cycle it becomes apparent why they must usually be constructed in series if a high degree of purification is to be achieved.
In such a filter used as a primary or 'roughing' filter the growth is
copious but rudimentary; the filter's operation corresponds with the
first, and part of the second, stage of the purification cycle, *i.e.*
E_1–E_2 E_3 in which a major part of the action is attributable to
adsorption of the substrates on to the film (sludge flocs). In the filter
the sludge is anchored to the medium instead of being suspended in
the waste, but otherwise the two processes are analogous. In the
rapid filtration process, however, the growth which becomes
detached is removed from the filter effluent by settlement and is
discarded; none of it is intentionally aerated and returned to the
filter. The resulting sludge is highly unstable, prone to become
anaerobic, and must be removed from the system as rapidly as
possible. It is usually very suitable for anaerobic digestion.

The settled effluent from the primary filter—30% to 40% of the
original BOD having been removed—then needs further purification
and this must necessarily be at a slower rate. The most readily
oxidised substrates—those with the lowest E_0—are preferentially
removed by the first filter, leaving the more intractable materials
to be dealt with in a second filter. However, the growth which
becomes established in the second filter should be of a different
nature, better adapted to oxidise the remaining impurities, so that
the overall rate of oxidation may be greater in two filters in series
than in a single unit attempting to do the whole job.

CHEMICAL TREATMENT

While considerable success has been achieved during the last 30 years in adapting and accelerating biological processes of waste purification, there has been hardly any fundamental new development since the introduction of the activated-sludge process. Attention has been focussed almost exclusively on the acceleration and control of natural degradation processes, and the question now arises as to whether this can ever be sufficient for the growing needs of industrialised societies. The effluent purification plant required for a chip-potato factory situated in an inland area provides a good illustration of the problem. With an intake of 500 tons of potatoes per day, the volume of effluent produced and requiring treatment is probably not less than 750 000 gal, even allowing for maximum practical re-use of water. The pollutional strength of this effluent will be at least three times as great as that of town sewage, which brings us to the equivalent of 2 250 000 gal of sewage of BOD 250 p.p.m. Yet the problem is essentially a chemical one—the removal of starch from water. Biological reactions are always comparatively slow and although they can be accelerated to some extent by increasing the concentration of the reacting mass and by elevation of the temperature, they cannot be expected to approach the velocities of chemical reactions. Some research schools are therefore giving attention to the possibilities of chemical methods of purification of waste waters and destruction of solid waste materials. This is by way of extension of the possibilities of chemical methods of treatment into fields hitherto dominated by biological processes. Generally speaking, it is already normal to tackle the purification of effluents of chemical origin primarily by chemical methods and these can be successful provided that the effluent can be kept separate from sewage until after treatment. Class (3) effluents are of this nature and the objective is to reduce their concentration of toxic substances to a level that may be disposed of safely by dilution only. Assessment of safe levels of concentration, however, is a constant problem.

INORGANIC TOXIC WASTES

The heavy metals which occur in natural waters are lead, copper, zinc, cadmium, chromium, and nickel. The first three of these are present naturally in the rivers and streams of some mining areas, and all six metals may be derived from electroplating wastes. Active mining operations may be expected to cause increases in the concentrations of lead, copper, and zinc in the areas where their minerals are mined and processed, owing to disturbance of the surface soil, dumping of waste, and the discharge of effluent waters from processing operations. Modern electroplating plant is designed to avoid wastage of the metal solutions used, but nevertheless some loss is unavoidable and traces of the metals are discharged in the effluents from such factories and eventually find their way, in some measure, into natural waters.

The discharge of plating wastes into town sewers is closely controlled in some countries, usually as a measure of protection for the biological sewage-purification plant in which the sewage is treated. Some of the metals combine readily with organic matter of biological origin, and an excess of chromium, in particular, has been known to inhibit the biological activity of purification plant. Whether or not damage is caused to biological treatment plant depends on the level of concentration of the metals present in the sewage. When the total concentration of lead, copper, zinc, cadmium, chromium, and nickel is less than 10 mg/litre it is doubtful if any inhibition has ever been demonstrated, and indeed it is possible that some of the metals are essential for the proper functioning of biological processes. Where no such plant exists, there is much less justification for strict control. In those circumstances where sewage treatment is restricted to physical methods only, the affinity of metals for organic matter is an advantage since their concentration in solution is considerably reduced as a result, without any harm to the treatment plant. The approximate extent to which sewage purification plant will remove these metals when they do not, together, amount to more than 10 mg/litre in the sewage treated is summarised in Table 2.1. These values are based largely on practical experience; as shown by Jenkins et al.[1] the solubilities of metals in a sewage depend on many factors, such as the pH value, the temperature, and the time of contact (from which one may suppose that the state of oxidation of the metal is involved).

When combined with organic matter, the metals become insoluble and are deposited in sedimentation tanks in treatment works, or in the bed of a stream or river. Whether organically combined metals are toxic seems doubtful in view of the well known occurrence of high concentrations of lead, arsenic and zinc, particularly, in shellfish and crustaceae; neither the fish nor their consumers appear to

TABLE 2.1

Removal of heavy metals by sewage treatment; 10 p.p.m. level

Metal	Sedimentation only (%)	Sedimentation followed by biological purification (%)
Lead	50	90
Copper	40	80
Zinc	60	80
Chromium	25	50
Nickel	30	30

suffer any ill-effects as a result. The concern of the present review is with the metals in solution.

Limits of acceptable concentrations of metals in waters to be used for water supply have been agreed internationally by WHO and these conform with the recommendations of the US Public Health Service.

These limits are in no sense toxic limits. The values were recommended for a variety of reasons. In regard to chromium, the Report states 'It is realised that the limiting value for chromium is well below the known toxic concentration, but it is considered that this element should not be present in drinking-water.'[2] One may infer from this that chromium was not thought objectionable in itself, but that it serves as an indicator of possible industrial pollution of the supply.

With the exception of lead, the various metals are more toxic, or harmful, to fish than to human beings. If, therefore, it is ensured that a water is not rendered unsuitable as a fishery by reason of its heavy metal content, its use as a source of water supply is also safeguarded, except in regard to its lead content.

TOXIC CONCENTRATIONS

The 'toxic concentration' of a metal cannot be expressed as a simple figure without limitations. Other factors, particularly the hardness of the water and the temperature, have strong influences upon it. This is because the toxicity of a substance depends upon its state of combination or dissociation. For example, cyanide ion is much less toxic than undissociated HCN so that the toxicity of a water containing cyanide is much greater at a lower pH value. Lloyd and

Herbert[3],[4] have reviewed the evidence for the modification of toxicities by chemical interactions of ionic species in waters and have pointed out that, in addition to pH value, dissolved oxygen concentration, carbon dioxide, and in certain cases sunlight have important effects which can be explained on physico-chemical grounds.

In order to express toxicities in single figures which can be compared with another it is necessary to restrict them to precise conditions of temperature and water hardness. 'Toxic concentration' also has to be defined. This is normally expressed in terms of the concentration in which one half of the test animals—usually fish— will die in a certain time. Thus '2-day TL' is the concentration which will be lethal in two days to 50 per cent of the organisms exposed. The 7-day TL value, which means 50 per cent survival for 7 days, is a figure which has been used in recent years, particularly by the workers at the Water Pollution Research Laboratory, Stevenage. This value, applied as a limiting concentration to river waters, would appear to be a 'safe' concentration especially in view of work done in Canada to which further reference will be made. Sprague and Ramsay[5] have used the Incipient Lethal Level, which is the concentration below which all the organisms survive indefinitely or, alternatively, 'that level of the lethal identity beyond which the organism can no longer survive for an indefinite period of time'. The Incipient Lethal Level is thus the TL value for infinite time.

The relationship between survival time and metal concentration is of the same kind for each of the metals: log (survival time) plotted against log (concentration) gives a curve which approaches the Incipient Lethal Level asymptotically. This means that the 7-day Toxic Level and the Incipient Lethal Level are the same for all practical purposes and accounts for the observation of Sprague and Ramsay[5] that 'usually 50 per cent mortality occurred within about 40 hours or not at all'.

The Incipient Lethal Level (ILL) is clearly the value which approaches most nearly to the 'Limiting Concentration' required for control purposes. It is apparent also that the ILL is largely unaffected by temperature,[3] which is a further advantageous feature. It is affected by water hardness, but in a predictable manner:

$$\log(\text{ILL}) = k + c \log (\text{total hardness})^{(4)}$$

Table 2.2 lists the ILL or TL7 values extracted from the literature,

TABLE 2.2

Incipient lethal levels of concentration to fish

Metal	*ILL (mg/litre)* 15 p.p.m. hardness	40 p.p.m. hardness	Fish (spp.)	Source
Lead	0·80	1·20	Rainbow trout	3
Copper	0·034	(0·095)	*Salmo salar* L.	5
	0·04	0·095	Rainbow trout	3
	(0·025)	(0·060)	Rainbow trout (yearlings)	8
Zinc	0·42	(0·90)	*Salmo salar* L.	5
	0·60	1·10	Rainbow trout	3
	(0·35)	(0·60)	Rainbow trout (3–4 months old)	8
Nickel	5·0	17·0	Rainbow trout	6
	0·56	(14·0)	Rainbow trout (14-day *Alevins*)	7
Cadmium	(0·003)	(0·008)	Rainbow trout and brown trout (14-day *Alevins*)	7

applicable to water of (a) 15 p.p.m. and (b) 40 p.p.m. total hardness (as $CaCO_3$). Figures enclosed in brackets have been corrected to the appropriate hardness using the data given by Lloyd.[4] The values for cadmium have been calculated on the assumption that the correction for hardness is the same.

ADDITION OF TOXICITIES

Investigations made by Lloyd in England and Sprague in Canada have proved that the toxicities of pairs of poisons are additive. The Water Pollution Research Laboratory at Stevenage has gone further and proposed that toxicities in excess of two are also additive.[9] There is sufficient evidence for this to make it acceptable as a working hypothesis. In general, where

$$\frac{A_s}{A_T} + \frac{B_s}{B_T} + \frac{C_s}{C_T} \cdots \frac{n_s}{n_T} = 1$$

the water will be potentially poisonous to fish. In this expression A_s, B_s, C_s, etc. are the actual concentrations of the poisons (in the present context, metals) present in the water, and A_T, B_T, C_T, etc. are the corresponding threshold values, or ILL concentrations. Thus referring to Table 2.2, a river water of 15 p.p.m. hardness containing 0·4 mg/litre of lead would not be toxic to fish, since $0·4/0·8 = 0·5$, but if it also contained 0·02 mg/litre of copper it would be toxic, since

$$\frac{0·4}{0·8} + \frac{0·02}{0·04} = 1 \ (= \text{one toxic unit})$$

Saunders and Sprague[10] have recently reported an extensive study of the behaviour of Atlantic salmon in avoiding a river polluted by zinc and copper as a result of mining operations. They found that adult salmon avoided entering the river when the combined concentrations of the metals reached 35 to 40% of a toxic unit, i.e. when $[Cu]/0·04 + [Zn]/0·5 = 0·35$. From the point of view of preventing fish kills, this avoidance reaction is an added safety factor in that if a discharge of toxic metals into a river was controlled so as to prevent the concentrations in the river exceeding one toxic unit, fish would not migrate from less polluted zones. It also means, however, that fish would not ascend the river and pass through the polluted zone to spawn in higher reaches.

It is unfortunate that no ILL value for chromium has been published. Statements have been made to the effect that hexavalent chromium is more toxic that trivalent chromium, but the various states of the metal are so readily inter-converted by oxidation and reduction that it is difficult to see how one may be tested for toxicity without the risk of its conversion to another. A review of the evidence for the toxicity of chromium[11] led only to the conclusion that much more work needs to be done. However, there have been no reports of toxic symptoms arising during the testing of concentrations of chromium such as are likely to occur in a natural water—of the order of 1 mg/litre—and Eden[12] suggested that 20 mg/litre could be acceptable in a river. That chromium is poisonous to cattle, as has sometimes been alleged, would seem to be refuted by the established practice of adding relatively enormous quantities of Cr_2O_3 to feeds in connection with digestibility studies.[13]

For the purpose of control of industrial discharges into streams it is suggested that the principle be adopted of limiting the concentrations allowable in the stream to the values shown in Table 2.3.

This would ensure that the water was not rendered unfit for drinking by the metals it contained. In any particular case the limits would be further adjusted so as to take account of the additive

TABLE 2.3
Suggested maximum concentrations allowable in streams

Metal	Concentration
Lead	0·05 mg/litre
Copper; Zinc	ILL appropriate to hardness
Cadmium	0·01 mg/litre
Nickel	5 mg/litre
Chromium	1 mg/litre

toxicities to fish and ensure that the total toxicity would not exceed one toxicity unit. Relaxation of the lead limit would be possible if the water was unlikely to be used for drinking purposes, when the allowable concentration could be based on the ILL values, taking into account the hardness and the other metals present.

ACCEPTABLE CONCENTRATIONS IN FACTORY EFFLUENTS

To arrive at acceptable concentrations of metals in the effluents from industrial premises, the concentrations already present in the stream must be determined, and the quantity which would bring these concentrations up to the acceptable limits, calculated. This requires a knowledge of the dry-weather flow (DWF) of the stream. For the present purpose, the dry-weather flow may be taken as the average flow during the times not immediately preceded by wet weather; it is not logical to attempt to cater for drought conditions. For example, if a stream naturally contains 0·02 mg/litre of lead and has a flow of 2 million gallons per day, the additional lead acceptable over a period of 24 h is 0·03 × 20 lb = 0·6 lb. If the discharge is to be made over a shorter period than 24 h, the quantity acceptable is proportionately less as we are concerned ultimately with concentration. A further allowance can be made if the volume of factory effluent adds considerably to the flow of the stream, but any other toxic constituents, or deficiencies, contributed by the effluent must also be taken into account.

Discharges into streams or rivers *via* municipal sewers should be controlled in a similar manner, but because of the effect of sewage in rendering metals insoluble, a further allowance may safely be made. For example; a river containing water of 40 p.p.m. hardness and having an estimated DWF of 10 m.g.d. receives 1 m.g.d. of sewage containing an industrial discharge of copper salts. The ILL for copper (Table 2.2) is 0·095 mg/litre, hence the discharge of sewage may be allowed to contain 0·095 × 11 mg/litre = 1·05 mg/litre or 10·5 lb per million gallons. As 40% of the inorganic copper will be rendered insoluble, the total allowable in the factory discharge is 17·5 lb per 24 h. In a case such as this, the desired conditions are more likely to be met if the factory discharge is made during working hours than if it were extended evenly over 24 h, owing to diurnal variation in the sewage flow. If the volume of industrial effluent in the above hypothetical case were 6000 gal/day, a limiting concentration of 300 mg/litre (approx.) of copper could be justified. This would, however, 'use up' the whole of the capacity of the sewerage system and the river to absorb copper, and in arriving at a final figure allowance would have to be made for the possible requirements of present and future industrial establishments in the town, and in other towns discharging into the same stream.

Although the problem is complex, the above example serves to show that Authorities have usually been unduly pessimistic in fixing 'standards' to be applied to industrial effluents containing heavy metals. When such an effluent finds its way directly or indirectly into a tidal estuary, or the sea, limiting concentrations in the receiving water would be difficult to justify and the only considerations to be taken into account are those relating to the proper operation of sewage treatment plant. Where no biological plant is in operation, no limits appear to be necessary as the metals are too valuable for any industry to throw away in such large quantities as to affect materially the concentration in the receiving body of water.

REFERENCES

1. Jenkins, S. H., Keight, D. G., and Ewins, Avril, *Int. J. Wat. Poll.* (1964), **8**, 679–93.
2. *International Standards for Drinking Water*, WHO, Geneva, 1958.
3. Lloyd, R., and Herbert, D. W. M., *J. Inst. Publ. Health Eng.* (1962), 132–45.
4. Lloyd, R. 'Biological Problems in Water Pollution', Third Seminar, 1962, Wat. Poll. Res. Lab., Stevenage, England.
5. Sprague, J. B., and Ramsay, A., *J. Fish. Res. Bd. Canada* (1965), **22**, 425–32.
6. *Water Polln. Res.*, 1965, p. 145, HMSO, London.
7. *Water Polln. Res.*, 1966, p. 52, HMSO, London.

8. Herbert, D. W. M., and Vandyke, J. M., *Ann. Appl. Biol.* (1964), **53,** 415–21.
9. *Water Polln. Res.*, 1965, p. 146, HMSO, London.
10. Saunders, R. L., and Sprague, J. B., *Water Res.* (1967), **1,** 419–32.
11. *Water Res. News*, 1960, No. 9, p. 16, Wat. Res. Assoc., Redhill, England.
12. Eden, G. E. 'Plating and Pickling Effluents', Seminar Lecture, Inst. Ind. Res. & Stds., Dublin, 1966.
13. *E.g.* Curran, M. K., Leaver, J. D., and Weston, E. W., *Animal Prodn.* (1967), **9,** 561–4.

Chapter 3

UNIT TREATMENT PLANTS FOR ISOLATED COMMUNITIES

E. H. NICOLL

INTRODUCTION

Over the last ten years or so various types of small activated-sludge sewage treatment plants have appeared on the market. Since many of the units are prefabricated or 'packaged' their attractiveness for dealing with the sewage from rural communities, individual establishments such as schools, hotels, and factories, or for providing temporary treatment facilities at construction sites, for example, is very real. At the same time there is a price to be paid for simplicity and the following remarks attempt to present a balanced view founded on experience gained since the first plant to appear in the United Kingdom was installed in Scotland in 1961.

The three principal types of completely mixed activated-sludge systems in common use are extended aeration, contact stabilisation, and another application of extended aeration, the oxidation ditch. In most designs primary settlement is dispensed with, so that it is raw unsettled sewage which is aerated in the presence of activated sludge. The various processes used are modifications of conventional activated-sludge treatment, and one important characteristic common to all types is that a very long period of aeration is provided at some stage to bring about oxidation of the sludge. It is emphasised that the purpose of aeration in such plants is not only to purify the raw waste, but by aerobic digestion or oxidation to reduce the amount of sludge produced. In the early days of 'packaged' extended-aeration plants this characteristic led to optimistic claims that there would be no net accumulation of sludge in plants and hence no surplus for disposal, but in practice this assumption proved to be erroneous. The reasons for this are that inert material present in the raw waste accumulates in the system, and that a proportion of the sludge produced is resistant to biological oxidation. In this context it may be amusing to note that experiments have been carried out in the USA on the disposal of human wastes within space vehicles. Aerobic conversion of the wastes resulted in a degradation-resistant fraction in the form of a brown pigment, which, for want of a better name, became known as 'hestianic acid', after the Greek deity, the virginal Hestia whose virtue remained resistant to the wooing of Apollo!

TREATMENT PROCESSES
Outline descriptions of the three systems already mentioned are given below. In general no primary settlement is provided, although screening or comminution may be incorporated ahead of aeration.

Extended Aeration
The extended-aeration process is illustrated in diagrammatic form in Fig. 3.1. The waste is passed in succession through aeration and

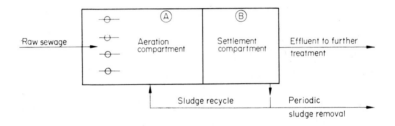

FLOW DIAGRAM (PLAN)

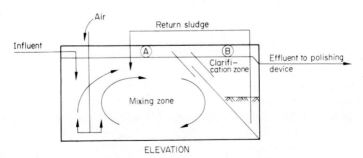

ELEVATION

Fig. 3.1   Diagrammatic form of extended-aeration process.

settlement compartments 'A' and 'B'. Aeration of the mixture of waste and activated sludge is for a period of at least 24 h based on dry-weather flow (DWF). It is again emphasised that the objective of this long period of aeration is not only oxidation of the suspended, colloidal and dissolved impurities, but also to achieve some aerobic digestion of the biological sludge produced. Following aeration, the mixture passes to the settlement compartment where the mixed

FIG. 3.2   Extended-aeration installation near Glasgow (by courtesy of Messrs Macleod & Miller (Engineers) Ltd, Blantyre).

liquor suspended solids (MLSS) are separated by gravity and are recycled to the aeration compartment. The clarified liquor flows out of the plant as effluent. Excess sludge is withdrawn periodically for disposal. An extended-aeration installation is shown in Fig. 3.2.

Contact Stabilisation
Contact stabilisation is illustrated in Fig. 3.3 and depends on the ability of healthy activated sludge quickly to adsorb and entrain polluting matter from raw sewage during a short period of up to about 2 h thorough mixing in the contact zone 'A'. From the contact zone the mixture passes into the central settlement compartment 'B'

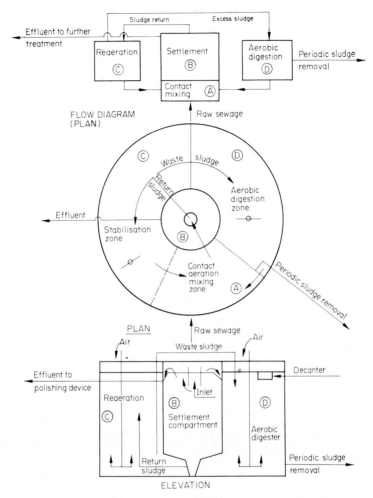

FIG. 3.3 Diagrammatic form of contact stabilisation process.

from which the clarified liquor is discharged as effluent. The sludge which settles out is passed to a re-aeration compartment 'C' where polluting matter is stabilised and further active biological organisms are produced, the principle being that the reconditioning of the sludge is more efficiently carried out after separation from the sewage. Eventually the sludge re-enters the contact zone so that the cycle becomes continuous. Excess activated sludge is wasted periodically

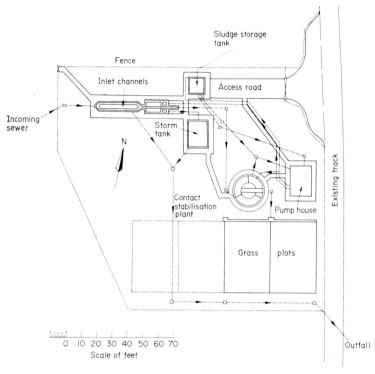

FIG. 3.4   Contact stabilisation installation serving Rossie Farm School, Angus (by courtesy of Messrs J. & A. Leslie & Reid, Consulting Civil Engineers, Edinburgh).

into an aerobic digester 'D' where it receives aeration over a prolonged period before disposal. A sewage treatment plant incorporating a stabilisation plant is shown in Figs. 3.4, 3.5 and 3.6.

Oxidation Ditch

Oxidation ditches consist essentially of racecourse-shaped continuous channels excavated in the ground. The type of installation most frequently found in the United Kingdom comprises a continuous flow ditch operated in conjunction with a detached final settlement tank as shown in Fig. 3.7. Aeration and circulation around the ditch are achieved by means of a rotor which maintains a velocity of flow sufficient to keep the activated-sludge flocs in suspension. As with the other systems, continuous recycling of settled sludge and a

FIG. 3.5 Contact stabilisation installation serving Rossie Farm School, Angus. Pumphouse and contact stabilisation unit on right, storm sewage tank centre background, inlet works centre foreground, sludge storage tank on left. (Photograph by courtesy of Messrs J. & A. Leslie & Reid, Consulting Civil Engineers, Edinburgh.)

FIG. 3.6 Contact stabilisation installation serving Rossie Farm School, Angus. Foreground, contact stabilisation plant; background, pumphouse. (Photograph by courtesy of Messrs J. & A. Leslie & Reid, Consulting Civil Engineers, Edinburgh.)

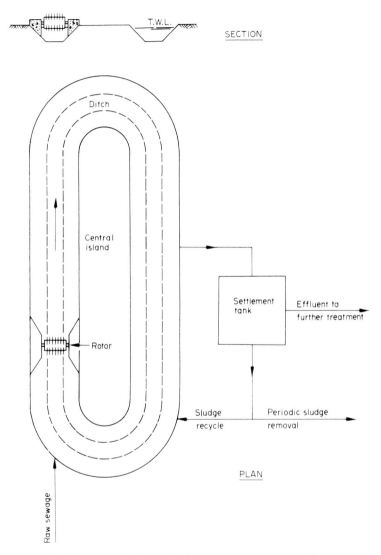

Fig. 3.7   Diagrammatic form of oxidation ditch—continuous flow type.

FIG. 3.8 Oxidation ditch at Glengoyne Distillery, Perthshire. Ditch showing rotor on left; sludge return arrangements and final settlement tank on right. (Photograph by courtesy of Messrs Cowan & Linn, Chartered Civil Engineers, Glasgow.)

FIG. 3.9 Oxidation ditch at Glengoyne Distillery, Perthshire. Foreground, final settlement tank; background, oxidation ditch. (Photograph by courtesy of Messrs Cowan & Linn, Chartered Civil Engineers, Glasgow.)

means of disposal of surplus sludge are necessary. Ditches are normally trapezoidal in cross-section and are usually designed to operate with a liquid depth of 3 to 5 ft. In certain cases where site conditions are suitable, partially lined ditches protected only in the vicinity of the rotor may be practicable, but more commonly they are wholly lined with concrete or rubber sheeting. Two oxidation ditch installations are illustrated in Figs. 3.8, 3.9, 3.10 and 3.11.

FIG. 3.10   Oxidation ditch at Glengoyne Distillery, Perthshire. Sludge return arrangements showing water wheel. (Photograph by courtesy of Messrs Cowan & Linn, Chartered Civil Engineers, Glasgow.)

BASIC CRITERIA APPLICABLE TO DESIGN AND OPERATION OF PLANTS

Owing to lack of information regarding extended-aeration plants, government departments in the United Kingdom adopted cautious attitudes towards them when they were first introduced. Difficulties were known to have occurred at certain installations, and between 1962 and 1964 an assessment of typical plants was made by the Water Pollution Research Laboratory. Subsequently, government departments judged it advisable to issue technical memoranda setting

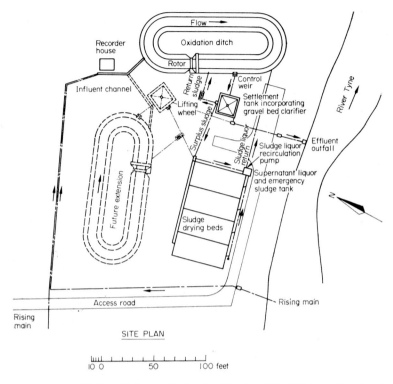

FIG. 3.11 Oxidation ditch installation for Burgh of East Linton, County of East Lothian (by courtesy of Messrs Robert H. Cuthbertson & Associates, Consulting Engineers, Edinburgh.)

out general provisions and design factors which would form an acceptable basis for the examination of schemes submitted for loan sanction or for consideration for grant. In the formulation of these criteria it was assumed that installations would be required to produce effluents of Royal Commission Standard, *i.e.*, suspended solids less than 30 p.p.m. and biochemical oxygen demand less than 20 p.p.m. The principal general provisions and design factors are given below and in Table 3.1.

GENERAL PROVISIONS

(i) The treatment process selected should be suitable for the particular waste which is to be purified.

(ii) The site of the treatment plant should be chosen to avoid the risk of annoyance from noise.

(iii) In general, surface water and subsoil water should be excluded from the sewerage system.

(iv) There should be adequate protection against corrosion.

(v) Standby electrical and mechanical equipment should be incorporated with provision for automatic changeover where practicable and for automatic restarting following power failure.

(vi) Arrangements should be made for the removal and disposal of surplus sludge at regular intervals.

(vii) Means of further treatment, *i.e.*, 'polishing' of plant effluent before final disposal, should be incorporated.

(viii) The provision of equipment for measuring and recording the flow to the plant should be considered.

(ix) Installation of prefabricated units must be carried out with care to ensure that protective coatings are not damaged, and the units must be set truly level.

(x) Skilled attendance is absolutely essential at regular and frequent intervals to ensure that installations are adequately operated and maintained.

In practice the success or failure of installations will depend to a large extent on the degree of attention paid to the basic criteria given above. Some of these will now be discussed in more detail.

Selection of Treatment Process
It is fundamental that the method of treatment should be suited to the particular waste concerned. Shock loads from the emptying of cesspools and chemical closets into sewerage systems have caused breakdowns in treatment at extended-aeration installations. Clay in subsoil water infiltrating into sewers and the presence of oil and grease can also adversely affect the activated sludge. It is not enough to install grease traps upstream of plants serving say hotels or restaurants; the traps must be emptied frequently.

The contact-stabilisation process is probably best suited to the treatment of sewage containing a relatively high proportion of its impurity in suspended and colloidal form as such matter should be readily adsorbed by the activated sludge in the contact zone, whereas organic impurity in solution is more likely to pass quickly through the compartment and hence out of the system.

Noise

Differing circumstances, including the relative sizes of plants, merit individual consideration, and it is difficult to be precise on the distance from houses which will avoid annoyance from noise. It may be that for small plants a minimum of 150 ft, or more if available, would suffice, but it is worth mentioning that the New York State Department of Health stipulates that a radial isolation of at least 250 ft should be provided for open extended-aeration installations. An extended-aeration plant in Scotland, designed to serve 1100 persons, is still audible at a distance of about 600 ft.

Surface Water and Subsoil Water

In general, surface and subsoil waters should be excluded from sewerage systems to which plants are connected. This is to say that units designed to treat foul sewage only, up to rates of three times dry-weather flow (3 DWF), would represent the ideal situation, but such straightforward cases do not always accord with conditions in practice. For example, even totally separate systems of drainage can be subject to such hazards as considerable flow variation, infiltration, illegal connections, connections made in error and imperfectly sealed open-ends left for future connections. Moreover, in practice, questions of whether to use such plants in conjunction with combined systems of drainage inevitably occur, and problems arise over the separation of storm sewage from dry-weather flows. It is not possible on information currently available to state a safe upper rate of flow in terms of multiples of DWF which could be sustained through the units without the risk of excessive or at least intermittent loss of suspended solids. Each case must be carefully considered having regard to its particular circumstances.

Corrosion

During erection the epoxy resin coatings must be protected from damage. Where packaged plants are set in the ground they are often provided with a system of cathodic protection. The sacrificial anode buried packs should be readily accessible and should be extracted periodically for inspection.

Standby Equipment

For plants making use of diffused air, provision should be made for the operator to be able to adjust the air supply to suit the requirements of a particular waste or any local circumstances which may apply. One method which allows considerable flexibility in the operation of the duty equipment and also ensures that a full supply of air can be made available should a duty machine ever require attention for any reason is to provide three blowers, two operational and one standby, each rated at 50% of the total air requirement. If three speeds are available, then by using one blower or a combination of two blowers and various pulley wheels, any one of eight outputs, being different proportions of the plant's aeration potential, can be achieved. Again, one of the two duty blowers can be controlled by a time switch set to cut it out for periods during the night when the load arriving at the works would be at its lowest and the dissolved oxygen content of the mixed liquor would be high.

Sludge Production and Desludging

There is no primary sludge to be dealt with. Moreover, the long periods of aeration involved should result in the surplus sludge being more stable than that produced by a conventional treatment works, and in a special sense, less in quantity. This is to say that it is the production of dry solids which is reduced, to perhaps 0·4–0·5 lb/lb of applied BOD, but the actual volume of the liquid sludge depends on its moisture content. Thus, if sludge were withdrawn at MLSS = 5000 p.p.m., or 99·5% moisture, then its volume would correspond to about 1·2 gal/hd d. To reduce the volume it is usual at extended-aeration installations to shut off the air supply for about an hour before desludging. If consolidation to say 98·5% moisture content could be achieved by this means, the volume of sludge would be reduced to about 0·4 gal/hd d. Relatively high concentrations of suspended solids can be carried in the aerobic digestion compartments of contact stabilisation plants.

Generally the periods between desludging of extended-aeration plants should not exceed four weeks, and it is important that the concentration of MLSS remaining should not be reduced below about 2000 p.p.m., since experience has shown that excessive withdrawal can have an adverse effect on subsequent performance. It may be worth considering the provision of sludge consolidation tanks to allow sludge to be withdrawn relatively frequently for storage and dewatering. Such tanks may be covered and so arranged that decantrate can be recirculated to the plant.

Effluent 'Polishing'

It has been found in practice that many plants are prone to intermittent discharges of high concentrations of suspended solids. Hence to safeguard the quality of final effluents means of final treatment of effluents before ultimate discharge are normally required. This is especially important for the smaller units which are more sensitive to variations in flow and where less supervision is likely to be provided than at larger installations. Irrigation over grass plots or passage through upward-flow gravel bed clarifiers are methods commonly adopted.

Performance Control

As aids to control, the provision of equipment for measuring the flow to the plant and the dissolved oxygen concentration in the mixed liquor are commended. The latter should exceed 1 p.p.m. at all times. At diffused air plants the incorporation of an air meter will enable the amount of air actually being supplied to be checked periodically.

Installation

A number of prefabricated plants have floated during installation, and some after erection. Flanged ports, located at the bases of tanks and subsequently blanked off, can be a useful preventative measure. Plants should be located at levels that ensure that neither surface water running off surrounding areas nor flood-water from streams can enter the units. Fences should be provided at an early stage to safeguard children and to protect the units.

Attendance

It is essential that plants are visited at daily intervals for maintenance purposes. Simple settling tests carried out on the mixed liquor and plant effluent can provide useful, albeit rough, visual indication to operators that all is not well and that further advice is required. It is important that air filters and aeration devices are kept clean to ensure an adequate and uniform supply of air. Plants must be desludged at regular intervals.

Aeration Requirements

It is not possible to be dogmatic about the amount of air which will result in successful operation of waste treatment. The design factors will ensure, however, that there is sufficient air potentially available for use if required, and provided that some flexibility in its application is allowed for, it should be practicable to arrive at the optimum air supply to suit the particular waste and circumstances involved. The design factors provide the potential for a well-nitrified effluent to be obtained.

Upward Flow Velocities

The surface overflow rate at maximum flow should not exceed $450 \, \mathrm{gal/ft^2}$ per day. This corresponds to a nominal upward velocity of $3 \, \mathrm{ft/h}$.

FLUCTUATION IN HYDRAULIC LOADING

Not enough attention has been paid in the past when designing sewage treatment works of various capacities to factors arising from differences of size involved, and it is emphasised that a minor works must not be regarded simply as a scaled-down version of larger units. Of prime importance for the plants under discussion are the effects of load variation, and especially fluctuation in hydraulic loading.

The types of situation prone to produce extremes in hydraulic loading occur when the contributory population is small, where the pattern of activity of individuals in the community is similar and where the treatment unit is situated close to the properties served. Units serving small groups of houses and establishments such as schools, hotels and factories may be especially at risk. In such situations a treatment unit designed to cope with a peak flow of say 3 DWF can be severely overtaxed hydraulically during certain periods of the day.

Research has been carried out in America on the performance of extended-aeration plants which were operated under different patterns of flow for a whole year. Whereas average BOD and suspended solids removals of 90% and 80% were obtained for all plants operating under steady flow and continuously varying flow, the performance figures dropped to 82% and 74% respectively for the pattern of flow typical of schools, that is where the load is discharged over only 8 h. In this situation the loss of efficiency was due

TABLE 3.1
Design factors
BOD of domestic sewage, assumed at 0·12 lb/hd d

No.	Item	Extended aeration	Contact stabilisation	Oxidation ditch
1.	BOD loading of aeration zone(s) should not exceed	15 lb/1 000 ft³ d	30 lb/1 000 ft³ d (contact + re-aeration)	13 lb/1 000 ft³ d
2.	Capacity of aeration zone(s) per population equivalent	8 ft³/hd	4 ft³/hd (contact + re-aeration)	9 ft³/hd
3.	Capacity of aerobic digester	—	3 ft³/hd	—
4.	Circulation velocity, not less than	—	—	1 ft/s
5.	Aeration requirements			
	(i) For diffused air systems, the duty equipment should be capable of producing at aerator depths of			
	(a) 6 ft	Up to 4 500 ft³/lb BOD	—	—
	(b) 9 ft	Up to 3 000 ft³/lb BOD	—	—
	(c) 12 ft	—	Up to 2 700 ft³/lb BOD	—
	(d) 15 ft	—	Up to 2 300 ft³/lb BOD	—
	(ii) For mechanically aerated systems, oxygenation capacity should not be less than	2 lb oxygen/lb (applied) BOD	—	2 lb oxygen/lb (applied) BOD
6.	Settlement compartment			
	(i) Overflow rate at maximum flow should not exceed	450 gal/ft² d	450 gal/ft² d	450 gal/ft² d
	(ii) Maximum nominal upward velocity	3·0 ft/h	3·0 ft/h	3·0 ft/h

to the intermittent but sustained high hydraulic loading which resulted in suspended solids being washed out of the system.

The foregoing underlines the necessity for an effluent 'polishing' stage following treatment and also the need to adopt conservative upward flow velocities. Where establishments such as schools, factories, etc., are concerned, the advisability of flow-balancing should be carefully considered.

CONCLUSION

There can be little doubt that small unit treatment plants have come to stay. They have certain advantages such as economy of head and that they may occupy less land. When operating properly they are odour-free and virtually foam-free. They can be useful for establishments and communities remote from main drainage facilities and may be used to provide temporary treatment pending the advent of main drainage. Sludge disposal and treatment are simplified as there is no primary sludge; digestion plant is not necessary and the sludge produced is readily dried and is odour-free. There are, however, certain disadvantages, which are essentially the price to be paid for simplicity. Plants are dependent on an uninterrupted supply of electricity, and mechanical failure may result in cessation of treatment. They require frequent and skilled maintenance. They have to be situated sufficiently far from habitation to avoid the risk of annoyance from noise. They are sensitive to high rates of flow and effluent 'polishing' is necessary in order to produce an effluent consistently complying with the Royal Commission Standard. In general, surface and subsoil waters should be excluded from the sewerage systems to which they are connected.

It is suggested, however, that such units have a part to play and that if designers and those responsible for operating the plants paid due regard to the points presented in this chapter, there would be sufficient flexibility built into installations to enable optimum performance to be sought for the varying wastes and circumstances likely to be encountered in practice.

ACKNOWLEDGEMENTS

The author's grateful thanks are accorded to the various firms who provided the plans and photographs of the sewage treatment installations in Scotland which serve to illustrate the text. The courtesy of the Scottish Development Department in agreeing to the publication of this chapter is also acknowledged. Opinions expressed are personal and are not necessarily those of any government department.

Chapter 4

CONVERSION OF SLUDGE INTO UTILISABLE GAS

R. W. BROOKS

On the face of it the title is specific, but I must make a small pedantic point—very little sludge is *converted* into gas and I suggest that perhaps an alternative title might be 'Digestion of Sludge to produce usable gas'. Digestion by *aerobic* bacteria in the presence of oxygen is a perfectly feasible process—although somewhat expensive—but it does not produce usable gas. So we are concerned with *anaerobic* digestion of sludge. I must also make it clear that this chapter concerns the digestion of *sewage* sludge. Digestion of many organic substances is possible and is in fact practised. These may include sludges from the treatment of organic industrial wastes and in some cases it is considered economic to use digestion methods for the treatment of whole waste liquors—usually when such liquors are very strong. (Slaughter-house, meat packing, etc.)

THE ANAEROBIC DIGESTION PROCESS
Anaerobic digestion is a process which occurs widely in nature, producing marsh gas in swamps and stagnant pools—the so-called 'will-o'-the-wisp'—and it has been defined as 'A biological process in which organic matter, in the absence of oxygen, is converted to methane and carbon dioxide.' For the moment I will not seek to define sewage sludge, but leave it as a general description covering primary sludge from sedimentation tanks together with whatever secondary sludges are produced in various bacteriological processes. Anaerobic digestion is generally considered to take place in two stages—acid formation and gas formation. In the first stage a group of micro-organisms breaks down carbohydrates, fats, and proteins

into simpler substances such as alcohols and volatile fatty acids. Some of these substances are used as food by the organisms and some persist in the sludge, but most of the lower volatile fatty acids (acetic, propionic, and butyric acids) will be converted to methane and carbon dioxide by a different group of organisms. This operation constitutes the second stage of the digestion process. The processes are in fact much more complex than this and many learned papers have been written on the subject, but for our purpose it is essential only to understand that there are two stages even though they become intermingled. The acid-forming organisms are much less sensitive than the methane-producing organisms and it is the latter which effectively control the whole process, since the two stages must stay in step with one another.

The next important question is *'Why digest sludge?'* Incidentally, we should not think of this as a recently adopted process; after all it has been going on in Imhoff tanks and septic tanks since the late nineteenth century.

Sludge disposal is the biggest single problem in the sewage treatment field. We know of various methods of treating sewage and we can make our choice according to the size of the works, effluent quality required, etc., but we shall still be left with sludge.

Digestion can help us here:

Firstly it reduces the weight of solid matter to be disposed of, by 30% or more.

Secondly digested sludge is an inoffensive material—it has a tarry smell which is not objectionable and thus one of the constraints on disposal methods is removed.

Thirdly some authorities believe that digested sludge dewaters much more easily than crude sludge and therefore sludge drying becomes easier. Others say that this is not so and that digested sludge may be less amenable to dewatering. Even so, the fact that it is inoffensive makes it possible to put the digested sludge on to drying beds without creating a nuisance. I have no personal experience of the comparable effectiveness of drying beds using crude and digested sludge so I will not comment. However, separation of water in secondary digesters and in deep lagoons (both of which are used in Coventry) is certainly feasible and relatively straightforward.

Fourthly digestion partly destroys pathogenic and parasitic organisms and thereby increases the potential uses of the sludge.

Fifthly digestion usually reduces the grease content of sludge and it thus becomes more attractive as a fertiliser.

Sixthly and last, but by no means least, a valuable end product, sludge gas, is produced.

So far as Coventry is concerned it is important that we should be able to dispose of our sludge in deep lagoons. A few years ago our digester capacity was inadequate and part of our sludge had to be pumped to the lagoons undigested. People living in the vicinity complained very strongly about the smell emanating from the lagoons—and their complaints were justified. We have now doubled our digester capacity and every gallon of sludge is properly digested before going to the lagoons. The local inhabitants have not merely stopped complaining—they have assured us that they cannot detect a smell at all. So far as sludge gas is concerned we are able to generate the whole of our power requirements from it, using the gas as a basic fuel with a small admixture of fuel oil.

The next important problem is *'How to digest sludge.'* It is possible to do this at ambient temperatures (so-called cold digestion) and perhaps for the very smallest plants this may be the only practicable method. In such a case a long retention period is necessary—probably 4 months—and the rate of production of sludge gas will scarcely justify collection.

At the opposite extreme is the thermophilic range at optimum temperatures around 60°C. Digestion is very rapid in this range, but heat loss is a severe problem and it appears likely that digestion at such temperatures will be somewhat unstable.

The generally preferred range is mesophilic with an optimum temperature around 35°C and this is the one normally referred to

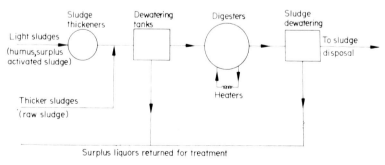

FIG. 4.1   Diagram of sludge digestion (heated) process.

when heated sludge digestion is quoted. It is represented diagrammatically by Fig. 4.1.

OPERATION OF SLUDGE DIGESTION

We should now look at the technical criteria governing anaerobic digestion. The basic aim should be to operate the process as consistently as possible and this applies to all the variables. I will discuss the practicalities later and for the moment merely outline the requirements.

Temperature should not be allowed to change significantly—there is good scientific reason for this since the activity of the microorganisms would be affected. There is an extremely good engineering reason too—the Coventry digesters in winter weather conditions may require upwards of 100 000 000 Btu's per day to heat the incoming sludge to 95°F and to maintain the temperature of the digester contents (4 000 000 gal) to compensate for heat losses. If the average temperature of the digesting sludge were allowed to fall off by 1°F then we should require an additional 40 000 000 Btu's to restore this loss. One can very soon be on a slippery slope since reduction of temperature brings reduced gas yield and a vicious spiral is soon set up.

Adequate mixing is important since newly introduced raw sludge must be brought as quickly as possible into intimate contact with the actively digesting sludge so that the organisms are able to work on the new sludge.

There is a natural tendency towards separation since the new sludge is normally at a lower temperature than the general digester contents and therefore tends to sink because its specific gravity is higher. This can bring further trouble as I will explain later. The evolution of gas will tend to create a certain amount of convection movement but, unless the tank is very deep relative to its plan area, this is unlikely to be sufficiently vigorous.

The digesters must have sufficient capacity—that is, one must provide an appropriate retention period. In general practice this will mean 25 to 30 days retention although it has been claimed that with thorough mixing this can be reduced to fewer than 20 days. This must always be a matter of degree since complete digestion would require an infinite time. An extremely important factor in calculating retention time is the water content of the sludge. If the water content of the digesting sludge is reduced from 96% to 95% then for the same amount of solids the volume of the liquid sludge is reduced by 20% and the retention increased by 25%. It is a standard saying that water does not produce gas and it is certainly very important that the water content of the sludge should be reduced as much as possible before feeding to the digester. This can of course be taken too far and it is necessary to consider the additional difficulty of adequately mixing a thick sludge. The best consistency of sludge will vary with cases, particularly with the mixing methods, but a feed sludge of

FIG. 4.2   Dewatering tanks: 45 ft long, 25 ft wide, 14 ft deep. Capacity of each 98 000 gal.

around 94% water content will usually be satisfactory. This may very well require a sludge thickening process prior to introduction to the digesters; normally, a simple dewatering process involving quiescent settlement and withdrawal of supernatant liquor. Such dewatering tanks will serve other purposes—allowing mixing of different sludges and acting as balancing tanks so that feeding to the digesters can be regular, and consistent. Illustrations of sludge dewatering and thickening tanks are shown in Figs. 4.2 and 4.3.

Another aspect of water content which might be forgotten is that heat is required to maintain the temperature. If too low a solids content sludge is fed it will be necessary to increase the turnover to achieve the same solids loading. We should therefore be heating water unnecessarily.

The rate of feed of organic matter should be regular and small compared to the digester volume. A charge once per day with a retention period of 25 days will amount to 4% of a digester volume and this is probably about as big a charge as one should use. It is thought by some authorities that charges should be much smaller

FIG. 4.3   Secondary sludge thickeners: 22 ft diameter, capacity 67 500 gal.

and there may be considerable advantage when a digester is in a sensitive state or if variation of the rate of gas yield is undesirable. We should note here that consistency of charging applies both to the volume and quality of the new sludge. It is no good measuring the volume of each charge with great accuracy if the solids content is varying from 6% to 2% since this would involve a 3 to 1 variation of charging rate of solid matter. Charging rate can be expressed in a number of ways. Some authorities regard the loading rate of organic matter as fundamental. This is only a different way of expressing loading since for a crude sludge of 6% solids, three-quarters of which are organic, a retention period of 25 to 30 days will involve a loading of about 0·1 lb organic matter per cubic foot of digester volume per day.

A further factor which tends to be forgotten is that consistency of withdrawal is also necessary. Obviously sludge must be withdrawn from a digester to make room for a new charge. One will withdraw marginally less than is put in since gas production will have slightly reduced the volume of the digesting sludge (by about 0·1 % per day —hardly noticeable, but 500 gal on an average digester). A more important point is that repeated withdrawals and charges must leave a digester in its standard condition. Gasification reduces the solids content of the digesting sludge—this will therefore be less than the solids content of the raw sludge. Take the case of a sludge fed to a digester at 4 % solids, three-quarters of which are volatile. If one-third of the volatile solids are converted to sludge gas and water this will reduce the solids content from 4 % to approximately 3 %. You will remember that earlier I said that there were further problems associated with raw sludge moving to the bottom of a digester. Withdrawal of sludge is often arranged from the bottom of the tank and it is possible to withdraw sludge that was only charged 24 h previously. This has many effects—all unpleasant. The water content of the digesting sludge rises, the efficacy of the micro-organisms falls off. Since part of the organic matter is withdrawn before it can be converted, the production of fatty acids and of gas is limited. The withdrawn sludge (which would be considered as *nominally* digested sludge) will smell offensively and sludge digestion will tend to become condemned.

To avoid these troubles it is imperative to ensure that average quality digested sludge is withdrawn and that the digester contents are kept constant. Of course if stirring is 100 % efficient then the sludge may be withdrawn without problem. Usually it is safer to arrange for withdrawal from different points and I would advocate the bottom of the tank and the liquid surface level. It is not advisable to avoid bottom withdrawal altogether since there will always be some accumulation of detritus (no matter how much care is taken with grit removal at the works inlet—and no matter how powerful the stirring) and occasional use of the bottom outlet will prevent the detritus from building up.

The next problem is the prevention or control of the build-up of scum on the liquid surface. Relatively light fibrous substances such as hair and textiles together with rubber and plastics materials float to the surface of the sludge and together with grease and oil form a thick mat of scum. There have been cases where such a scum was

6 ft thick—in a 30-ft-deep digester. This takes up 20 % of the volume and decreases the retention time. Once again really efficient mixing will reduce the problem, but it is safer to provide mixing *and* a surface draw off. If scum build-up is not controlled, sooner or later it will be necessary to enter the digester and remove the scum physically.

It is important that the pH of a digester should be maintained at around 7·0 and it is also important, in order to ensure sufficient buffering capacity, that alkalinity should be maintained at a high level. Unfortunately the scientists do not agree what the level should be, neither do they offer much in the way of suggestions as to how to maintain pH and alkalinity. One or two people have suggested adding lime, but great care must be taken with the mixing. I know of no better method of clogging sludge pipelines. If a digester is operated carefully and consistently the pH and alkalinity will be satisfactory anyway. The same applies to the volatile acids level which is also suggested as an important parameter. Change in volatile acids is a useful indicator, but usually the only significant controls which can be exercised are the rates of feed and of with-drawal—bearing in mind that this means control over solids in the digester. In my view a more sensitive control parameter indicating the state of the digester is the quality and quantity of the gas yield. Changes in the proportion of methane and carbon dioxide are usually very rapid and significantly meaningful. The significance may vary from works to works but a 1 % increase of CO_2 and corresponding decrease of CH_4 should be regarded with suspicion.

DESIGN OF DIGESTERS

Now we can move on to the practicalities of digestion and see how the operational parameters are provided for in the design of the digesters and ancillary equipment. As a background to these points the population of Coventry is around 350 000 with a dry weather flow of 24 million gal/day almost all of which is treated at one works. The sedimentation tanks are circular and are desludged at intervals of 2 to 4 h. The bacteriological processes are arranged in three treatment streams—one of these is alternating double filtration while the other two involve two-stage treatment—partial activated sludge followed by filtration on bacteria beds.

There is an average production of sludge of about 200 000 gal/day, 80 % of this being primary sludge and the remainder thickened humus sludge. The mixture has an average solids content of about 3 to $3\frac{1}{2}\%$. Surplus activated sludge is returned to the sedimentation tanks and this has so far created no problems.

I have already mentioned sludge thickening prior to charging digesters—at Coventry we use simple rectangular tanks equipped with floating arm draw-offs for the removal of supernatant liquor or top water. This draw-off takes liquid from just below the surface thus avoiding the scum which forms on the top. There are six tanks each with a capacity of 100 000 gal. The amount of supernatant liquor taken off may vary considerably—but is normally around about 40% of the total sludge, thus increasing the solids content to about 6%.

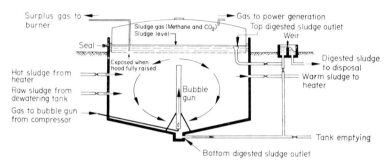

FIG. 4.4   Diagram of sludge digestion tank incorporating stirring device.

The digesters themselves (Fig. 4.4) are circular reinforced concrete tanks 60 ft in diameter. There are two different cross sections—the four original tanks (constructed in the early 1950s) have a side water depth of 21 ft and a conical floor with a 30-degree slope. The four later tanks (built in 1967) have a side water depth of 30 ft and the floor has a slope to the centre of only $11\frac{1}{4}$ degrees. The maximum depth of both types of tank is approximately 36 ft. The volumes of the two types of tanks are respectively 450 000 and 550 000 gallons per tank, giving a total volume of 4 million gal. Each tank is fitted with a Firth Blakeley Sons & Co. Ltd floating steel gas hood having a maximum vertical movement of 10 ft and a working range of 8 ft. The hoods throw a pressure of 6 in of water gauge. The eight hoods have a combined storage capacity, between top and bottom working levels, of approximately 200 000 ft^3 or just over half a day's gas yield. There are differences in detailed design, but in both cases the hoods have a side clearance of 6 in and operate with spiral guide rails running between fixed rollers. They are also equipped with vacuum and excess pressure relief valves and fitted with Lea Recorder Co. Ltd level indicators giving remote indication in the adjacent power house (Fig. 4.5).

FIG. 4.5   Sludge digesters: 60 ft diameter, 36 ft deep. Capacity of each 450 000 gal.

Withdrawal pipes are provided at the surface as well as at the bottom of the tank—both are valved and connected to an external weir box so that no deliberate act of sludge withdrawal is necessary. One of the two valves is opened and the other closed and the act of charging sludge automatically ensures an appropriate discharge over the weir. The discharged sludge runs to a sump whence it is pumped automatically by a bucket pump to the sludge disposal works some 3 miles away. The normal arrangement at the present time is for the valves to be set so that 6 days in the week withdrawal is from the surface and the seventh day from the bottom of the tank. For tank emptying purposes it is possible to valve off the weir chamber and to connect the bottom withdrawal pipe direct to the suction side of a bucket pump. Each digester is equipped with a feed pipe to deliver new sludge from the dewatering tanks, and suction and delivery pipes

FIG. 4.6   Gas fired sludge heaters.

connected to external sludge heaters (Fig. 4.6). All the sludge mains are arranged in an underground pipe gallery thus providing access for valve operation and maintenance. Heat insulation is provided by constructing the digesters partly below ground and by placing excavated material in a bank round the tanks (Fig. 4.7).

There are a few interesting points in the construction of the digesters. Four digesters and two dewatering tanks were built in the 1950s, but by 1965 these were grossly overloaded and work commenced on an extension to provide four more digesters and four more dewatering tanks. At the same time it was decided to provide a gallery for all the sludge pipes and to carry out modifications to the existing digesters. This was a fairly complex project, but it was made much more difficult by the necessity of maintaining the existing digesters in operation whilst the new ones were being built and then to commission the new digesters so that the old ones could be taken off stream for modification. The site is rather restricted and the contractor had a difficult time. Incidentally, this was the first contract on which we had specified placing concrete in a 30-ft-high wall by pouring full height in a purpose made shutter. This is commonplace today but at that time many engineers were specifying pours not exceeding 9 ft or even 6 ft in height.

FIG. 4.7   Sludge digesters—heat insulated by means of earth embankments.

Two of our digesters are fitted with gas circulating or stirring devices. We have an Aero Hydraulics Ltd bubble gun mounted at the centre of one of the tanks. This is a device which vaguely resembles an 18 in diameter air lift, except of course that it operates on compressed sludge gas. The main difference from a conventional air lift is that the gas is fed into a syphon which releases large bubbles at intervals. These bubbles fill the tube and travel upwards like a piston displacing sludge in front and behind. The bubbler unit is timed so that one bubble is released as the previous one reaches the surface. This arrangement improves the efficiency very considerably and a compressor of 3 hp is adequate to stir the contents of the digester sufficiently to keep scum formation to a minimum. The second digester has a different arrangement of gas circulation and that will be mentioned later. The remaining six digesters are not at present stirred other than by sludge circulated through the external heaters. The effect can only be slight due to the enormous inertia involved—in any case each digester is on circuit for only 6 h out of 24. Before the old digesters were converted and provided with surface draw-offs there were considerable difficulties with scum build-up. We have had little recent trouble, but I believe that it is too risky to operate on a permanent basis without stirring.

In our present practice, dewatered sludge is charged twice per 24 hours to each of the digesters. We have experimented with multiple charging and could in theory charge each digester eight times a day—a total of 64 charges. However, we have not been able to detect any significant improvement in gas yield or digester efficiency during a period of multiple sludge charging. It is our earnest wish that one day sludge handling, from sedimentation tanks to the dewatering tanks to the digesters to disposal, may be done automatically. If and when this comes to pass we shall conduct much longer experiments since we feel that 'little and often' ought to show benefits. Probably the ideal arrangement would be such that the peak rate of gas production coincided with maximum gas demand—but this is obviously not simple to arrange. For the time being we have fitted some of the sluice valves with mechanical actuators and these can be operated from the sludge pumping station.

The sludge gas produced consists of about 65% methane and 33% of carbon dioxide with the remainder a mixture of hydrogen, nitrogen, hydrogen sulphide, and various hydrocarbons. It has a nett calorific value of 580 Btu/ft^3 and the average rate of production is 360 000 ft^3 per day or just over 1 ft^3 per head of population per day. The equivalent fuel oil would have a value of £24 000 per annum and therefore good use must be made of this gas. The system at Coventry is to use the gas as the main fuel in English Electric Co. Ltd dual-fuel engines (Fig. 4.8). These are basically modified diesel engines and since ignition is by compression a minimum amount of fuel oil must be used. This is about 10% of the fuel oil requirement at full load and does not vary through the load range, fuel input being controlled by a governor operating on the gas and air supplies. During 1968–69 approximately 84% of the energy used was provided by sludge gas. The engines drive alternators which generate electric power for the works. At the present time the demand averages 800 kW and this is generated and distributed on a ring main at 6600 V and transformed to 415 V at the centres of demand.

The dual-fuel engines are water-cooled and the jacket water is circulated through sludge heaters and heat exchangers. Sludge from each of the digesters in turn is pumped through Ames Crosta Mills sludge heaters where heat is taken up from the jacket water and the sludge maintained at 35°C. In the event of there being more heat to be dissipated from the engines than is required by the sludge the heat exchangers transfer heat to a secondary water circuit of cooling water pumped from and returned to the adjacent river. In the event of more heat being required, boilers on the engines' exhausts can be brought into circuit.

There are two sludge circulation circuits each linking four digesters with two sludge heaters and equipped with three sludge pumps. In the event of breakdown the two circuits can be interlinked. The

FIG. 4.8   Power house: 8-cylinder, 600 hp dual-fuel engines, 418 kW generators, 6600 V.

standard arrangement is for each digester to be 'on circuit' for 6 h out of 24, but in the event of any discrepancy in sludge temperature this can be varied. During a normal 6-h period 144 000 gal or about 30% of the digester contents will be pumped from each digester, through the heaters and back to the same digester. Control of sludge heating is exercised at the power house including operation of the motorised sluice valves controlling the circulation on each digester.

If gas demand and supply do not match, some buffer capacity is provided by the volume of the gas hoods. After that has been used then either the gas must be burned to waste if there is excess, or if there is a shortage then one of the dual-fuel engines must be run on straight diesel oil for a period until gas stock is restored.

There are many alternative ways of carrying out sludge digestion, and certainly very many ways of handling sludge disposal. Various manufacturers are involved and naturally advantages are claimed for each arrangement. Digesters may be fitted with solid covers instead of floating hoods—in that case gas storage is provided in a separate gas holder. This will certainly resolve the problem of floating hoods not moving in unison. Sludge heating may be internal and a compromise solution involves heat transfer in heat exchangers fitted onto the tank wall. Stirring may be carried out by pumps suspended from a fixed or floating hood—gas circulation may be by means of various devices and may involve different amounts of gas.

Simon Hartley Ltd have developed their Heatamix unit, both for stirring sludge and heating the digesting sludge. Hot water can be supplied to the unit and heat can be transferred to sludge which passes up the tube (again somewhat similar to an air lift). The company claim gas yields 20% higher than average—presumably due to the energetic mixing. Each unit is rated at 50 ft^3/min of gas throughput and they claim the sludge turnover to be double the gas usage. There are two types of Heatamix unit—one type is mounted inside the digester. This is the arrangement at Coventry, where the units do not incorporate water jackets since sludge heating is carried out externally as already described. The two units are very effective and the stirring is good, but the gas compressor in use in this case is 11 hp. Incidentally, a point of interest at Coventry is that there appear to be greater heat losses from the two digesters equipped with stirrers. This does not invalidate the mixing, but draws attention to the necessity for adequate insulation. The alternative arrangement of Heatamix units is to mount them externally—this has obvious advantages and the effectiveness of stirring is apparently just as good.

Ames Crosta Mills & Co. Ltd have been developing plant for sludge digestion for many years—including mechanical stirring devices for both fixed roof and floating gas hoods. Dorr Oliver & Co. Ltd and other manufacturers have also produced machinery for sludge digestion.

Sludge is fed at about 15 000 gal/day in two charges from the dewatering tanks to each digester at an average solids content of about 6%—about three-quarters of which is organic. Over 50% of the organics are destroyed during an average retention of just over 30 days and each digester yields around 45 000 ft^3 of gas per day. This is equivalent to about 15 ft^3/lb of organic matter destroyed, or about 8 ft^3/lb of organics added.

HEAT EXCHANGERS

Depending on the degree of hardness of local town water supplies, all sludge heat exchangers suffer from scale deposition on the sludge contact side of the exchanger. In areas with hard water supplies frequent (12 months upwards) maintenance may be necessary to maintain adequate rates of heat transfer. Annular cylindrical heat exchangers with relatively narrow sludge spacings also suffer from choking with foreign matter and may require rodding at not more than weekly intervals. (These are possibly cheaper in capital cost than tubular heaters, but are an example of increased revenue maintenance for less capital expenditure.)

SAFETY

Adequate purging of digester spaces and gas lines is essential during commissioning and withdrawal of units if explosion risk is to be reduced to a minimum. Care should also be taken at the design stage to ensure that any leakage of gas into buildings or galleries cannot build-up to the explosive range without dispersal or detection. (Lower explosive limit of sludge gas is approximately 5% in air mixture. Higher explosive limit of sludge gas is approximately 15% in air mixture.)

There are two principle ways of guarding against gas accumulation.

(a) Plenty of natural or forced ventilation.
(b) Detector devices operating automatic alarms. These are delicate devices in practice requiring frequent calibration and checking in order to give a workable reliability. Ventilation is much to be preferred.

GAS METERS

It has been our experience that the installation of multiple gas meters can only be an approximate guide to the performance of an engine or digester. Wide fluctuations in balance between gas produced and gas used have been experienced, all of which cannot be explained by pressure and temperature differences at the metering positions. This is still being investigated. The meters used at Finham are the rotary vane type supplied by Parkinson Cowan.

Another point worth mentioning is the starting up of a digester. This is best done by using digesting sludge from another digester if this is available. For a new installation such seed sludge is not available and other methods have to be used. The digester is either *filled* with crude sewage and the sewage replaced over a period by daily feeds of sludge, or the daily sludge make is transferred to the digester gradually filling it. Use of the first method permits the operating temperature to be attained earlier than the second method, but a longer time is required to obtain a suitable solids content. A period of at least 12 to 18 months would be required to obtain a solids content equivalent to 1 lb of organic matter per cubic foot of digester volume.

GENERAL UTILITY OF SLUDGE DIGESTION

Of recent years a campaign against sludge digestion seems to have developed. There are problems, of course, but they can normally be overcome. However, in a few instances digestion has been inhibited or even stopped altogether. Because there was a great deal of talk about digestion troubles the Water Pollution Research Laboratory (supported by the Institute of Water Pollution Control, the Ministry of Housing and Local Government and the Scottish Development Department) conducted a country-wide enquiry. This involved the issue of a large number of questionnaires—some 1400 were returned and although the findings have not been published formally, a first assessment has been presented at Institute meetings by Dr Swanwick, and his colleagues.* Even this 'first assessment' is a somewhat bulky document, but a few of the salient points are as follows.

It seems that nearly 50% of the population of England, Scotland, and Wales is served by treatment works incorporating sludge digesters. It is known that there are at least 323 works, serving a population of over 21 million, operating sludge digestion. Some other returns were not complete and I shall refer only to those where the forms have been completed in all respects. There are 142 works using heated digestion, used in areas ranging in population from over 500 000 to below 5000 and serving in total nearly 19 million people. Fifty-one of these works reported some difficulties with sludge digestion, but five-sixths of these were temporary troubles.

* Swanwick, J. D., Shurben, D. G., and Jackson, S., *Wat. Pollut. Control* (1969), **68**, 639–61.

Of all the cases of difficulties reported, 62 % were due to inadequate design or operation and these were categorised in order of importance based upon the population served as follows:

(1) Stratification and loss of solids
(2) Overloading
(3) Temperature control
(4) Scum
(5) Poor mixing
(6) Accumulation of grit
(7) Unspecified
(8) Pump blockage
(9) Mechanical breakdown.

Therefore, provided the digesters are properly designed and constructed, are adequate in capacity and are operated properly there is little to fear in these cases.

In a further 32 % of the cases trade wastes were considered to be the cause of the difficulties and here the order of importance, based upon population served, was quoted as

(1) Toxic metals (plating, pickle liquor, etc.)
(2) General trade and chemical wastes
(3) Liquors from production of town gas
(4) Chlorinated organic compounds
(5) Formaldehyde
(6) Pharmaceuticals
(7) Tanning waste liquors.

The report gives little information under this section, but does quote concentrations of zinc of 0·7, 0·8, 1·6 and 3 % in dry solids. It would seem that trade effluent control was not very effective in some cases, but one does not know the full stories. Shock doses of toxic metal are likely to have much worse results than a fairly constant dose. The combination of various substances may be harmful too— the Water Pollution Research Laboratory suggest that the effect of metals may be strongly influenced by the concentration of sulphide. Perhaps we have been lucky at Coventry, but the only digestion trouble known to be associated with trade effluent was due to an accidental discharge of a slug of ammonia liquor from the local gas works.

The third category of difficulties (accounting for the remaining 6 %) was anionic detergents. It seems that there are possibilities of trouble if the concentration (as manoxol OT) reaches around 2 % of dry solids. Even then it is quite likely that difficulties will arise only when starting up a digester or where there are other coincidental problems such as overloading. It appears that in any case an answer has been found by WPRL—addition of stearine amine has been found to be extremely helpful.

COSTS

I was asked to include information on costs. This is relatively simple for capital costs of plant, but may not mean very much. An extension scheme completed 2 years ago included four digesters with gas hoods; four dewatering tanks, extensive pipe galleries, humus sludge thickeners. The total cost of this scheme was £238 000.

Operating costs are a great deal more difficult. Any sensible comparison would have to quote costs for alternative processes to take the place of our sludge disposal involving thickening, digestion and lagooning. This is not possible without a complete appraisal of the city's sludge disposal system. The operation of the digestion process, including dewatering, involves a running cost of about £10 000 pa—mostly labour—since this plant, like the rest of the works, is manned on a 24-h basis.

The economics of power production are a study in themselves and there are many considerations to be taken into account. The power requirements vary from one treatment works to another—obviously if the demand is far in excess of that obtainable from gas then works generation is unlikely to be economic. If, on the other hand, the proportion of power requiring straight diesel generation is small in total units, but high in maximum demand, then the Electricity Board will be at a disadvantage. Each case must be considered on its merits—obviously the total power requirement must be high enough to justify the manning of the power station and it may be possible to provide other local duties for the attendant.

A true assessment is difficult at Coventry because, among other things, the power house is designed for a higher electricity demand than has yet been required. The demand will increase during the next 12 months and it will then be possible to make a better assessment. On the face of it, however, we can improve on the Electricity Board's charges whilst having the comfort of a multiple supply.

SUMMARY

(a) Heated sludge digestion is an eminently practicable process which is widely used in England, Scotland and Wales.

(b) It can be extremely beneficial and has the fundamental advantage that there is a wealth of practical experience of the process.

(c) The quoted disadvantages are almost all derived from insufficient capacity, poor design, or incorrect operation and any sludge disposal or sewage treatment process would suffer under these circumstances.

Chapter 5

TREATMENT OF MEAT TRADE EFFLUENTS

M. C. DART

As with most industries, there is little difficulty in disposing of the waste waters from slaughterhouses or packing-houses in large towns where there are adequate facilities for the treatment of sewage, but in rural areas problems sometimes exist. The object of this chapter is to present information which will enable the magnitude of the problem to be assessed approximately and to describe briefly the methods which may be used for its solution.

SOURCES OF POLLUTION

The chief sources of polluting matter in waste waters from slaughter-houses are: (1) faeces and urine; (2) blood; (3) washings from carcases, floors and utensils; (4) undigested food from the paunches of slaughtered animals; (5) domestic sewage from canteens and toilets and, sometimes; (6) condensate from the rendering of offal. In meat processing factories there may also be wastes from the cooking, curing, and pickling of meat.

Blood contributes considerably to the polluting load, and at the larger slaughterhouses it is usually collected for separate disposal by coagulation and drying. If efficient collection of blood is not carried out the polluting load from a slaughterhouse may be increased by as much as 40 %. It is often stated that recovery of blood is no longer economically practicable, but if the cost of treating the additional organic load imposed by discharging blood with the effluent is considered then this is not usually found to be true.

75

POLLUTING LOAD
Strength and Volume
The amount of polluting matter in the waste waters from meat processing factories obviously depends on the number of beasts killed and the nature of the processes. Before designing any waste treatment process it is essential to have reliable figures for the volume and composition of the waste waters. Not only are average values required but also some indication of the maximum values of flow rate and strength. Where an effluent flow recorder is installed there is little difficulty in determining the volume discharged daily and the maximum rate of flow. If an effluent flow recorder is not available then the volumes can sometimes be estimated from water meter readings. To determine the composition of the effluent, 24-h composite samples, preferably taken over a full working week, should be analysed. Snap samples seldom give information of any value because of the wide and rapid variation in composition which can take place. Trade effluent surveys should preferably be carried out at periods of maximum production, but in any case the results should be related to production figures.

TABLE 5.1
The polluting load and volume of effluent discharged from some slaughterhouses in Britain

	Slaughterhouse				
	A	*B*	*C*[a]	*D*	*E*
Types of beasts	Pigs	Cattle Pigs Sheep	Pigs	Cattle Pigs Sheep	Broiler chickens
BOD load					
(lb/hog-unit)	1·3–1·8	1·1	0·45	1·2	—
(lb/1 000 lb live weight)	6·5–9·0	—	2·25	—	9–11
Volume of effluent					
(gal/hog unit)	87–91	80	31	38	—
(gal/1 000 lb live weight)	435–455	—	155	—	825–1 525

[a] Figures shown under 'C' are only for the killing of beasts. Those shown under 'A', 'B', 'D' and 'E' are for killing and processing.

It is sometimes necessary to calculate an approximate polluting load in order to make preliminary assessment of the size of treatment plant required. This can be obtained from published figures relating polluting load to number of beasts killed and processed. In the earlier American literature it was the practice to assume that the polluting load produced by killing one fully grown cow or bull was about 2·5 times that produced by killing one pig, sheep or calf, and polluting loads were related to the number of 'hog units'. The more recent practice is to relate polluting load to the live weight of beasts killed. Estimates of the polluting load and volume of waste water, per hog unit and per 1000 lb live weight of beasts killed, from several British slaughterhouses and meat packing plants are given in Table 5.1. These may be compared with values of 6·5 to 23·5 lb of 5-day BOD/ 1000 lb live weight (mean 15·0 lb 5-day BOD/1000 lb) and 625– 3620 gal/1000 lb (mean 1470 gal/1000 lb) reported by the US Department of Health[1] for American plants. It is interesting to note that the polluting load from killing and packing chickens corresponds roughly with that from the slaughtering of pigs. In small rural slaughterhouses the water consumption is often much less than in the larger modern plants, but the trend is towards larger factories with continuous killing and butchering lines and a relatively high water consumption.

Reduction of Polluting Load at Source

It is often possible to make considerable reductions in polluting load by good housekeeping within the factory, or by modifications to the process. For example, a slaughterhouse which operates an efficient blood recovery system will probably give rise to 40% less polluting load than a slaughterhouse of similar size which allows the blood to flow to waste. In order to pick out the stages in a process where a reduction in polluting load can be made it is often valuable to carry out a wastes survey within the factory. The general results of such a survey carried out in a poultry packing station are given in Table 5.2.

After examining the figures given in Table 5.2 it was decided that the stages where significant reductions in polluting load could be made most conveniently were the transport of feathers and waste offal. Both these operations were carried out by a constant flow of water, and by substituting dry disposal a reduction of up to 25% in the polluting load would be possible. In making process changes such as these there is usually a conflict of interest. On the one hand a

TABLE 5.2

Polluting character of individual waste waters from the killing and packing of broiler chickens

Origin	5-day BOD (p.p.m.)	4-h permanganate value (p.p.m.)	Estimated volume	5-day BOD load (lb/day)
1. Crate washing	900	280	2 800	25
2. Hot dip and wing scald	1 560	570	1 080	17
3. Feather flume	1 825	330	1 000	18
4. Waste offal flume	2 640	215	1 500	40
5. Edible offal flume	78	16	1 000	1
6. Wash spray	4 025	320	420	17
7. First chill	780	78	5 400	42
8. Second chill	790	113	5 500	43
9. Washdown	2 440	370	1 000	24
10. Domestic	400	—	1 050	4
Total			20 750	231

reduction in polluting load leads to a reduction in the capital and operating costs of the treatment plant, but against this must be set the possibility of lowering the bacteriological quality of the product. In any food industry of course the produce quality must always be the first consideration.

PRETREATMENT

Waste waters from killing and butchering operations usually contain a high proportion of coarse suspended matter, whilst those from processing operations contain less suspended matter but considerable quantities of fat. In modern killing and processing plants the two streams are often kept separate for pretreatment.

Screening

In order to keep the polluting load reaching the treatment plant to a minimum it is essential to screen out as much of the coarse solids from the killing and butchering line as possible. This is most usually carried out on fine-mesh screens ($\frac{1}{8}$ in to $\frac{1}{4}$ in aperture) of the vibrating, rotating, or mechanically cleaned type.

Grease Removal

Grease can be removed from the processing waste waters by conventional grease traps which bring about the coagulation of fats by cooling followed by separation of the solid or semi-solid material in baffled chambers from which it is removed by skimming. Dissolved air flotation is finding an increasing application in the United States for the removal of grease.[2]

In the latter process air is dissolved, under pressure, in a portion of effluent from the flotation unit. This effluent is then mixed with the waste waters to be treated, and as the pressure returns to atmospheric the dissolved air comes out of solution, forming bubbles on the particles of grease or solids. Each particle is thus attached to a bubble of air which lifts it to the surface of the liquid, where it can be skimmed off. Dissolved air flotation units have a short retention time (20–30 min), so that grease and fat are separated quickly before there is time for acidity to develop. This has the advantage that a higher quality grease is recovered and that the effluent passing to the treatment plant is not acidic.

Balancing

The rate of discharge and the strength of waste waters from slaughterhouse meat-processing plants are extremely variable, and in order to make efficient use of any treatment plant it is usually advisable to provide a balancing tank large enough to even out the flow of waste waters over a 24-h period. In order to estimate the smallest balancing tank that will be satisfactory it is necessary to know full details of the variation of flow or strength throughout the day. Often this information is not available, and in this case it is usual to provide a balancing tank with a capacity of about two-thirds of the daily flow. Provision should always be made for the removal of any sludge which settles out in the balancing tank.

PRIMARY SETTLEMENT

Where the waste waters are to be treated by percolating filters it is usual to provide primary settlement to reduce the suspended solids content. The size of settlement tanks used varies widely but Imhoff tanks with retentions of 1–3 h have been used in the United States and are reported to remove about 65% of the suspended solids and 35% of the 5-day BOD.[1] In Britain a settlement tank with a retention time of 35 h was reported to remove 39% of the 5-day BOD when treating waste waters from the slaughtering of pigs.[3]

Where a conventional settlement tank is used a retention time of about 4 h would seem to be sufficient. A more important design criterion is the 'Surface Overflow Rate'. Under normal conditions this should be less than 900 gal/ft^2/day at maximum flow in order to bring about effective removal of suspended solids.

In cases where both balancing of flow and settlement are required some economy in capital cost can be made by carrying out both functions in the same tank.

Where the waste is to receive further treatment by the anaerobic biological process primary settlement is not necessary. If the waste is to be treated by the activated-sludge process primary settlement is not strictly necessary but is sometimes provided.

CHEMICAL TREATMENT

The addition of coagulants such as 'alumino-ferric' to the waste waters prior to settlement greatly increases the removal of organic matter. At one poultry packing station in Britian addition of alumino-ferric equivalent to 17 p.p.m. of aluminium, followed by settlement for 1 h, reduced the 5-day BOD of the waste waters from 856 p.p.m. to 305 p.p.m. (64% reduction). The addition of coagulants is not usually favoured because of difficulty of disposing of the large volumes of sludge produced.

ANAEROBIC BIOLOGICAL TREATMENT

Considerable reductions in the 5-day BOD of slaughterhouse wastes can be brought about by anaerobic digestion. This method has been used in the United States,[4] New Zealand,[5] and more recently in the United Kingdom.

This process differs from the anaerobic digestion of sewage sludge in that a relatively short retention time is used (24 h) and the anaerobic organisms are removed from the treated liquid by settlement and returned to the digester. The process is thus analogous to the activated-sludge process. Organic material in the waste is broken down to methane and carbon dioxide and the mixture of gases is collected and burned to maintain the digester contents at the required temperature. The reaction can be represented in a much simplified form by the equation:

Organic matter + micro-organisms →
more micro-organisms + CH_4 + CO_2 + energy

An investigation using a pilot scale anaerobic digester with a capacity of 500 gal for the treatment of slaughterhouse waste is reported by the Water Pollution Research Laboratory.[6] This plant was operated at a loading of 0·1 lb of 5-day BOD/ft^3 of digester volume per day and the contents were maintained at 33°C. The concentration of volatile suspended solids in the mixed liquor was maintained at about 10 000 p.p.m. Under these conditions a 95% reduction in BOD was obtained. One difficulty reported during the operation of the pilot plant was the loss of anaerobic sludge in the effluent due to bubbles of gas causing the particles of sludge to float to the surface of the settlement tank. This was attributed to the digestion process continuing after discharge from the digester, and was overcome by stirring the liquor to disengage gas bubbles before settlement and by fitting the settlement tank with a spiral baffle.

Recently, two full-scale anaerobic digestion plants, designed to treat slaughterhouse and meat processing waste waters, have been installed in the British Isles. It is understood that both these plants have experienced operating difficulties due to loss of suspended solids in the effluent.

This problem has been overcome on the full scale, at least partially, by blowing air through the digester liquor before settlement. This has the double effect of inhibiting the action of the anaerobic organisms and disengaging the gas bubbles from the sludge particles.

Effluent from an anaerobic digestion plant will not normally be suitable for discharge to an inland watercourse without a further stage of aerobic biological treatment.

OXIDATION PONDS

An oxidation pond is the simplest form of aerobic biological treatment and can be regarded as bringing about the natural purification processes occurring in a river in a more restricted area.

Oxidation ponds are used in countries where land is cheap and weather conditions are favourable. In the United States typical oxidation ponds are about 4 ft deep and have retention times of 30 to 120 days. The oxygen required for the growth of the aerobic organisms is provided partly by transfer across the air/water interface and partly by algae under the influence of sunlight.

The environment in an oxidation pond is not easily controlled, mainly due to lack of mixing, and organic material can settle out near the inlet to the pond causing anaerobic conditions and offensive smells. Overloading of the pond also causes anaerobic conditions since the oxygen demand is greater than can be supplied by the natural oxygenation processes.

Provided that oxidation ponds are properly designed and maintained and weather conditions are favourable, they provide an economical form of waste treatment. However, the climate in Britain generally prevents them from operating satisfactorily except under exceptional circumstances.

PERCOLATING FILTERS

In a percolating filter the aerobic micro-organisms exist in a slime or film which is supported on the surface of the filter medium. The waste is applied to the surface of the filter and trickles down through it whilst air percolates upwards through the medium and supplies the oxygen required for purification. As the organisms are fed the amount of biological film increases and some of it breaks away. This must be removed by settlement in humus tanks before the effluent is discharged.

One of the problems encountered in the treatment of strong organic wastes on percolating filters is that of 'ponding'. This is due to excessive growth of biological film which blocks the surface of the filter. In order to overcome this difficulty it is usual to dilute the waste before application to the filter by recirculating final effluent and also to use alternating double filtration.

With alternating double filtration two filters are used in series— the effluent from the first filter being applied to the second. The first, primary, filter is loaded at a relatively high rate and so begins to develop a heavy growth of film.

The secondary filter which is treating the partially purified effluent from the primary filter is lightly loaded and the micro-organisms in the film are starved. After a short period, usually a week, the order of the filters is changed and this alternating light and heavy loading prevents the build-up of the excessive film which causes ponding.

If recirculation and alternating double filtration are employed, percolating filters provide a satisfactory method of treating waste waters from the meat industry, and effluents of a high standard can be obtained. Results reported by the Water Pollution Research Laboratory[7] indicate that if the waste waters were screened, settled, diluted 1:1 with recirculated effluent and treated by alternating double filtration at an overall loading of 0·585 lb of 5-day BOD/ yd^3 day an effluent containing 11 p.p.m. of 5-day BOD and 25 p.p.m. of suspended solids was obtained. The effluent from the primary filter was not settled before application to the secondary filter.

In America, the use of high rate single-stage percolating filters is common, and any tendency of the filters to pond is controlled by the use of a high recirculation ratio and large (4-in) filter medium. One such filter, treating slaughterhouse wastes after preliminary treatment in a septic tank, is described by the US Department of Health.[1] With a filter loading of 0·54–0·77 lb of 5-day BOD/yd^3 medium per day and a recirculation ratio of about 5:1 the overall removal of BOD after final settlement was reported to be 92–98%. The successful use of a plastics filter medium in a pilot scale plant treating slaughterhouse wastes has been described by Garrison and Geppert.[8] They reported that at a loading of up to 16·6 lb of 5-day BOD/yd^3 medium per day and a recirculation ratio of 2:1 about 71% of 5-day BOD was removed. The relatively high temperature of the waste applied to the filters (30–33°C) was probably the reason for the successful operation of the filters at such high loadings.

ACTIVATED SLUDGE

In the activated-sludge process the waste waters are mixed with a suspension of aerobic micro-organisms (activated sludge) and aerated. After aeration the 'mixed liquor' passes to a settlement tank where the activated sludge settles and is returned to the beginning of the process to treat more waste. The supernatant liquid in the settlement tank is discharged as the plant effluent. Air can be supplied to the plant by a variety of means, including:

(a) blowing air into the mixed liquor through diffusers;
(b) mechanical surface aeration;
(c) allowing air from a sparge pipe to be broken up into fine bubbles by a rotating impeller which also provides the turbulence required to keep the sludge in suspension.

All the methods are satisfactory provided they are properly

designed to give the required concentration of dissolved oxygen in the mixed liquor (greater than 0·5 p.p.m.) and maintain the sludge in suspension.

Activated-sludge plants generally have a lower capital cost than percolating filters, and even when allowance is made for the higher running costs they are usually economically more attractive. As with all biological systems there is a production of surplus micro-organisms and this surplus activated sludge must be disposed of, but some activated-sludge plants are operated without primary settlement and so avoid the production of offensive primary sludge. In some modifications surplus sludge is stabilised by further aeration.

One activated-sludge plant at a poultry packing station in East Anglia employs an 'Oxigest' unit to treat approximately 60 000 gal of waste waters per day with a 5-day BOD of 800 p.p.m. Screened waste waters are passed to a balancing tank and then to aeration tanks operating at a loading of approximately 23 lb of 5-day BOD/ 1000 ft^3/day. Air is supplied by Roots type rotary blowers through diffusers. Analysis of twenty-two samples of effluent taken during the period 6 September 1965 to 5 January 1966 had an average 5-day BOD of 18·5 p.p.m. and an average suspended solids content of 21 p.p.m.

The treated effluent has to be discharged to a substantially dry ditch and so it is 'polished' by means of a grass plot with an area of 720 yd^2. After the grass plot the effluent to the ditch had an average 5-day BOD of 12 p.p.m. and an average suspended solids content of 9 p.p.m.

Another type of waste which is amenable to treatment by the activated-sludge process is that derived from the manufacture of gelatine. Pilot scale studies over a three-month period show that at loadings of up to 30 lb of 5-day BOD/1000 ft^3/day an effluent of 'Royal Commission Standard' (20 p.p.m. 5-day BOD and 30 p.p.m. solids in suspension) can be produced. The main purpose of this pilot scale work was to establish the maximum loading at which the unit would produce an effluent with a methylene-blue stability of 5 days. The desired stability was obtained at loadings of up to 40 lb of 5-day BOD/1000 ft^3/day.

TERTIARY TREATMENT

The quality of the final effluent from properly designed biological processes is often governed by the performance of the final settling tank. In practice settling tanks with overflow rates of between 600 and 900 gal/ft^2 day at maximum flow will give effluents having suspended solids concentration in the region of 20 to 30 p.p.m. Provided that the biological process has been carried out efficiently the BOD associated with this concentration of solids will be in the range of 12–20 p.p.m. The most economical method of achieving effluents of consistently better quality than this is to use a further tertiary treatment or 'polishing' stage. This is usually more economical than increasing the size of the biological plant or settlement tank.

A variety of methods has been used for tertiary treatment of effluents and some of these are described briefly in the following paragraphs.

Grass Irrigation Plots
This method of tertiary treatment is a very economical one and is convenient where removal of 60 to 70% of the suspended solids is required. Loadings of from 100 000 to 750 000 gal/acre day may be used and the plots require little maintenance except for occasional cutting to control weed growth.[9] It is advisable to provide at least two plots so that one can be allowed to dry out for cutting.

Upward Flow Gravel Clarifiers (Banks Type Clarifiers)
Work by Banks[10,11] has shown that removal of between 70 and 90% of the suspended solids can be achieved by passing the effluent up through a 6-in bed of $\frac{1}{4}$-in pea gravel. Loadings of up to 2500 gal/yd^2 day may be used. These clarifiers are reported to require little maintenance beyond back washing at intervals of from 4 to 11 weeks.

Slow Sand Filters
Filtration of the effluent through a bed of sand at rates of up to 270 gal/yd^2 day has been used to give solids removals of 60 to 70%.[9] At these low loadings back washing is not required but the filter must be allowed to dry out periodically so that solids can be removed from the surface of the filter together with a small amount of sand. For this reason it is essential to provide at least two sand filters.

Rapid Sand Filters
Solids reductions of 70 to 90% can be achieved by filtration through sand at rates of 20 000 to 40 000 gal/yd^2 day.[12] At these high

filtration rates, the head loss increases rapidly to a value of 8 or 9 ft. These highly loaded filters require back washing at daily intervals, and again more than one unit should be provided to maintain continuity of treatment.

Microstraining
Microstrainers originally developed for water-works use are finding an increasing number of applications in the polishing of effluents. At loadings of up to 50 000 gal/yd^2 day solids removal of between 55 and 70% has been reported. This method of tertiary treatment tends to be rather costly but may be justified on larger installations.

REFERENCES
1. US Department of Health Education and Welfare. *An Industrial Waste Guide to the Meat Industry*. US Public Health Service Publication No. 386, 1958.
2. Kalinske, A. A., 'Flotation in waste treatment', In *Biological Treatment of Sewage and Industrial Wastes*. Vol. II *Anaerobic Digestion and Solids–Liquid Separation*, Ed. McCabe, J. and Eckenfelder, W. W., Reinhold, New York; Chapman and Hall, London, 1958.
3. *Water Pollution Research* 1960. HMSO, London, 1961, 37.
4. Steffin, A. J., 'Treatment of packing house waste by anaerobic digestion', In *Biological Treatment of Sewage and Industrial Wastes*. Vol. II *Anaerobic Digestion and Solids–Liquid Separation*, Ed. McCabe, J. and Eckenfelder, W. W., Reinhold, New York; Chapman and Hall, London, 1958.
5. Hicks, R., *Treatment of Auckland Meat Wastes*. Auckland Metropolitan Drainage Board, 1954.
6. Hemens, J., and Shurben, D. G. 'Anaerobic digestion of waste waters from slaughter-houses', *Food Trade Review* (1959), **29** (7), 2.
7. *Water Pollution Research* 1961. HMSO, London, 1962, 37.
8. Garrison, K. M., and Geppert, R. J. Packing house waste processing. Applied improvement of conventional methods. *Proc.* 15 *industr. Waste Conf. Purdue Univ. Engng. Extn.*, Ser No. 106, 1960, 207.
9. Truesdale, G. A., Birkbeck, A. E., and Shaw, D. A. 'Critical examination of some methods of further treatment of effluents from percolating filters', *J. Inst. Sew. Purif.* (1964) (1), 81.
10. Banks, D. H. *Surveyor* (London 1964), **123**, No. 3745, 21.
11. Banks, D. H. *Surveyor* (London 1965), **125**, No. 3789, 45.
12. Pettet, A. E. J., Collett, W. F., and Waddington, J. I., *J. Inst. Sew. Purif.* (1951) (2), 195.

Chapter 6

WASTES FROM FOOD MANUFACTURING INDUSTRIES

DENIS DICKINSON

BIODEGRADABLE WASTES

The food industry is universal; it is not even restricted to man, as certain animals store food, while communal insects not only store it but transform it as well. Every geographical area, therefore, has a food waste disposal problem which, indigenously, is roughly proportional to its sewage problem. If the area prepares food for export in addition to its own requirements, its waste disposal problem becomes greater. The waste is of the same nature as the food itself—vegetable matters, meats, and fats—all of it essentially decomposable by natural agencies. During transformation in a food factory—of which the domestic kitchen is the smallest unit—the food materials are cut, bruised, washed, and generally broken down physically, thus releasing enzymes and exposing cell contents to enzymic and bacterial action, so that food-plant wastes are essentially unstable. The natural substances they contain are utilisable almost completely by micro-organisms, the enzymes and micro-organisms capable of effecting their stabilisation are universally present and the natural processes will commence spontaneously and continue at a rate determined largely by temperature. It is the control of this process and the avoidance of incidental nuisances which pose a problem, particularly when a large amount of factory effluent has to be purified in a comparatively small space in order to avoid pollution of a nearby water or land area.

In general, industrial methods of food transformation are similar to domestic methods and as a population uses industrially prepared foods to an increasing extent the amount of waste from its domestic kitchens is likely to decrease. For example, as the use of prepared vegetables increases—ready-peeled potatoes, frozen chips, peas, and carrots—so the housewife must produce less effluent and garbage in her own kitchen. However, there are some differences resulting from the scale of operations; industry will generally produce less solid waste and more effluent in the preparation of a food. Where difficulties arise from food-manufacturing wastes it is generally as a result of their being concentrated in a particular drainage area, not because of the nature of the wastes themselves.

LIGHTLY-POLLUTED WATERS

Certain widely used processes give rise to large volumes of only lightly-polluted effluents. Vegetable washing is one of these. Root vegetables in particular are not extensively cut or damaged when they are washed to remove soil. This applies to potatoes and carrots which are washed prior to packaging; to beet, sugar-beet, and chicory root which are washed on receipt at the factory as a preliminary to processing. Wash water for this purpose is often taken from a stream or well and is recirculated several times after screening and settlement. The washing process may be combined with a water-transport or fluming system. Although the treatment of these used waters to fit them for re-use is a simple matter, it is not possible to state in general terms the size or type of screen required, nor the period of settlement necessary to remove suspended solids. This is because the screening operation most suitable depends on the mechanics of the water recirculation system employed, while the period of settlement which should and can be used depends on the quantity and properties of the soil adhering to the raw roots. Simple sedimentation and storage tests will soon show how easy or difficult it is to settle out the suspended matter and how long the settled soil will remain 'sweet'; these properties vary from one area to another.

Some washing operations are of only short annual duration and require essentially portable plant. The washing and plugging of strawberries is such a process and is best done in the growing area. The quantity of water used should not be very large, but because the fruit is comparatively soft and exudes juice the wash water becomes more polluted and should not be recirculated. It will pollute streams and is best disposed of by spraying on to land. As the fruit ripens during the summer over a short period there is usually no difficulty in arranging for the disposal of the effluent in this way.

The washing of semi-prepared vegetables, such as vined peas, is frequently not a simple operation: it may incorporate cooling of the material and also transport. Water carriage of fruits and vegetables in factories is a method which has wide applications, but it does use large volumes of water which becomes contaminated with vegetable matter and requires treatment before discharge or re-use. Fluming is widely used in fruit canneries, especially those processing peaches, pears, apricots, and grapes. The disadvantages of fluming outweigh its advantages wherever waste treatment and disposal are a problem, and alternative methods of transport should be considered. Unlike root vegetable wash-waters, fluming water for semi-prepared fruits and vegetables becomes contaminated with juices and is liable to undergo rapid biological deterioration. If it is to be re-used it must first be stabilised. In pea canneries and freezing plants stabilisation is achieved by chlorination to a level of 2 to 3 p.p.m. residual Cl_2, but this process is unsuitable for fruit transport waters. The National Canners Association (NCA) of the USA has reported considerable success by acidifying transport waters with citric acid to a pH value of 4·0; in handling peaches they saved 75 % of the volume of waste transport waters as a result of acidification and re-use. At the same time there was a 30 % reduction in the quantity of BOD discharged.

CONCENTRATED WASTE WATERS

The largest single potential source of BOD in all vegetable processing industries is the blanching operation. This is the stage in which the prepared vegetables are immersed in boiling water for a period of 1 to 10 min. During this immersion soluble solids diffuse from the vegetable into the hot water and starches swell. If the scale of operations is small this process is carried out batch-wise, and the blanching water is discharged after each batch. Thus the maximum volume is wasted but its BOD *concentration* is minimal, although the *quantity* of BOD discharged over a period is the maximum for the process. Most usually the blanching operation is carried out in a continuous machine from which there is a small continuous overflow during operation; the whole water content of the blancher will normally be discharged periodically. Continuous operation results in the quantity of BOD per ton of vegetable blanched being the minimum possible (assuming the blanching time and temperature to be optimum for the product), but its *concentration* may be very high indeed. Moreover, it is normal practice to empty the blancher completely at the end of a day or shift, which results in a massive discharge of highly concentrated waste water entering the factory drains during a very short period; and this water is very hot. Obviously some provision must be made to balance the flow of effluent before it reaches any treatment plant and preferably before it leaves the factory. It should be pointed out that the apparent remedy of allowing the blancher contents to cool before emptying and then running out slowly over a long period is *not* to be recommended as it would lead to contamination of the product by thermophilic bacteria whose growth within the machine would be favoured by such a procedure. If the factory is large enough to use a number of blanchers simultaneously, then it would be prudent to ensure that they are not all emptied at the same time, but it will still be necessary in most cases to provide balancing tanks of some kind before the effluent leaves the premises.

Vegetables and fruits which require peeling give rise to effluents containing high concentrations of suspended solids and colloidal matter which consists of vegetable substances in a raw or partially cooked state depending on the process used. The simplest effluent from the point of view of treatment is that produced by a cold-peeling process in which the peel is removed from the fruits or vegetables by abrasion or knives. Mechanical knives are used mainly for fruit such as apples and pears; most of the peel is then in a solid strip state and comparatively little of it need enter the effluent, especially if it is transported away from the peeling machines by mechanical or pneumatic conveyors. Fluming in water to some central point for screening, however, facilitates the transfer of soluble and fine suspended matters from the peel fragments into the water used for their transport, which acquires a high BOD in consequence. However efficient the subsequent screening operation may be, a sizeable BOD load will result from the use of fluming for the conveyance of peelings.

Abrasive peelers use water sprays to remove the fragments of peel from the fruit or vegetable and from the machines, so that fluming of the waste is an integral part of the operation. Because the fragments of peel and flesh are small they yield a high proportion of their soluble contents to the transport water before they reach the screens.

CAUSTIC PEELING

Caustic (or lye) peeling is a process favoured for the preparation of peaches, potatoes, and some other vegetables. It involves the immersion of the material in a bath of hot caustic soda—which may be from 2 to 25 % concentration—for a time sufficient to soften the skin, followed by passing through powerful water sprays and, for root vegetables, an abrasive machine in addition. The process combines the peeling and blanching operations and it is unusual to submit lye-peeled vegetables to an additional blanching process. Although lye peeling constitutes a considerable improvement to the processor, from the point of view of effluent treatment and disposal it is little less than a disaster. The peel and flesh removed are so transformed as to render them more soluble and colloidal; the effluent BOD is raised about tenfold, and the suspended solids content increased to a similar extent. Screening becomes virtually ineffective as a process for reducing BOD and even removal of the larger solids is made much more difficult. The caustic bath has to be removed from time to time; it is very concentrated and contains a high concentration of sodium carbonate (to which much of the original NaOH is converted during use) and must usually be dumped in some safe place to which it is transported by tanker. Biological purification of this material after appropriate dilution is certainly possible, but seldom economic.

It should be noted that all root vegetables which are to be peeled *must* be washed before they enter the peeling process. This will reduce wear on mechanical or abrasive peeling machines and will avoid the production of 'cooked mud' in any hot peeling process. This material can be very difficult to remove from an effluent by the normal process of sedimentation. If soil enters a lye bath, colloidal silicates may well result and even though most will remain in the lye bath, they will also enter the factory effluent by being dragged out on the vegetable surfaces, from which they are subsequently washed.

STEAM PEELING

In the process of steam peeling the peel is first loosened by heating in high-pressure steam and then removed by high-pressure jets of water. The effluent is highly charged with fragments of peel and flesh and has a high BOD. The proportion of the vegetable entering the effluent, however, is generally less than in other methods of peeling so that the substitution of steam peelers for other processes may result in an overall reduction in the total weight of BOD discharged

from a factory. Experience shows, however, that any such improvement is temporary as the introduction of steam peelers usually
increases the production capacity of a plant.

The increased attention given to effluent problems in recent years
has led to investigations of established processes with a view to
reducing their contributions to effluent strength and volume. New
methods of peeling have been devised. Great savings are possible by
conversion from fluming to mechanical/pneumatic conveyance of
both peelings and peeled product when the peeling or trimming
operation is essentially mechanical. A remarkable achievement of
this nature was described by G. Perkins concerning the preparation
and handling of artichoke products. A tenfold reduction in effluent
quantity was achieved in spite of greatly increased production.

DRY-CAUSTIC PEELING PROCESS
A modified lye-peeling process was developed at the US Department
of Agriculture's Western Region Laboratory, Albany, California,
and has been adopted quite widely for potatoes. The process involves
a short immersion of the potatoes in caustic soda of about 15%
concentration at 170°F (77°C) using a screw-conveyor and tube
system. After air cooling on a wire mesh belt conveyor, the potatoes
are tumbled in a rotating drum fitted with gas-fired radiant heaters,
when much of the peel is removed as a dry powder (Fig. 6.1). The

FIG. 6.1   Pilot-scale infra-red peeler, dry-caustic process, at the Western Regional
Laboratory, USDA, Albany, California.

FIG. 6.2   Peel removal unit, dry-caustic process, showing rubber-fingered rollers.

rest of the peel and the eyes are removed by passing between rubber rollers which have rubber fingers (Fig. 6.2). The debris is washed away with water sprays. The wastes from the tumbler drum are dry; those from the rubber-roller machine may contain as much as 23% solid matter and can therefore be handled as dry solid. The illustrations are from photographs of the pilot plant developed in the laboratory.

This process has become known as 'dry caustic peeling', and further developments have been made to adapt it for application to tree fruits, particularly peaches and apricots. It was necessary to design a different type of peel removing unit, and flexible rubber discs are now used instead of the fingers developed for the potato peeler. Factory trials have been run co-operatively with the NCA and commercial canneries and have shown savings of about 95% in effluent volume and 90% in terms of COD. The product yields from this process of peeling are at least equal to those from conventional processes, and the quality of both potato and fruit products is satisfactory.

INFLUENCE OF CAREFUL HANDLING ON EFFLUENT STRENGTH

Once the solid materials have been prepared for packing and processing—whether this involves canning, bottling, dehydration, pulping, freezing, or any specialised purpose—they are from two to three times as valuable as the raw material brought into the factory. The aim should be, therefore, to achieve as near as possible to 100% transfer into the final package; but many manufacturers are remarkably prodigal with these prepared materials. They tolerate inefficient filling and transport operations which permit unnecessarily large quantities to fall on to the floor and then into the factory drains. The contribution of spillage losses to the strength of factory effluents is not sufficiently appreciated. A quartered apple, for example, weighing about 40 g, if spilled on to the floor, subsequently squashed and then washed into the drains will contribute about 8 g of BOD to the effluent, which is equivalent to 2000 p.p.m. BOD if contained in a gallon of water. Similarly, each 1 lb of jam spilled, dripped from a filling nozzle, or otherwise lost to the drains contributes 0·7 lb of BOD to the effluent. There are therefore two distinct sets of reasons why losses of prepared materials should be avoided—both of them economic.

When the product contains both solid and liquid—such as canned fruits and vegetables in brine or sauce—two filling operations are involved and these require constant attention. Over-filling of cans with fruit causes losses of the syrup added subsequently; this occurs when the lid is sealed on the can or jar. Once a can is sealed no further loss of contents can occur, but this is not true of glass jars which are subsequently heat-processed. If there is insufficient room for the thermal expansion of the contents of a jar, the excess will be vented *via* the cap or lid into the water tank or steam chamber in which the processing is carried out and will then discharge with the factory effluent. Each gallon of 30 degree Brix syrup lost to the drains will contribute about 3 lb to the BOD.

There are, of course, unavoidable losses and accidental spillage of materials. Whenever possible these should not be washed into the drains but collected by brush and shovel or a vacuum cleaner. Continual dry-cleaning of operational areas in a food factory may do more to reduce the volume and strength of the factory effluent than any other single preventive measure. In addition it contributes to safety and spotlights inefficient machines or operations.

In modern factories the water used for cooling processed containers is recirculated after cooling and chlorination. Because of breakages of glass containers the cooling water needs to be replaced completely at intervals. An alternative cooling method employs the principle of evaporative cooling and passes the hot jars through a cold mist rather than through sprays or a mass of water. Much water can be saved in this way.

SEASONAL AND YEAR-ROUND OPERATIONS

Food manufacturing industries are very diverse and the size of factories varies tremendously. Factories may be classified as follows:

(1) *Seasonal operations only.* Examples: canning, drying and freezing of fruits such as apples, apricots, peaches, pears, and pineapples; tomato products; pulp, juice, powder; sugar manufacture from cane or beet; seafoods preparation and freezing.

(2) *Year-round operations with seasonal peaks.* Examples: dairy products—concentrated and dried milk, whey, chocolate crumb, cheese; vegetable canning, freezing, drying, and brining.

(3) *Year-round operations.* Generally products which are not made entirely or mainly from fresh or frozen ingredients.

Some of the most difficult problems of effluent treatment and disposal arise from factories in the first group. The scale of operations is usually very large and yet may continue for only a few weeks in each year. Kefford has reported an instance of a small town in Australia where it would have been necessary to increase the size of the municipal treatment plant by a factor of ten to accommodate the effluents from local fruit canneries which operate for a period of only 4 to 6 weeks during the year. Such problems are by no means unusual and in the past unsatisfactory disposal methods have been tolerated in sparsely populated areas where the people are largely dependent on the continued existence of the factories. The situation is changing, however, partly on account of increasing public interest in pollution abatement, but also because the factories continue to increase the scale of their operations to a point where the physical removal of effluent and wastes becomes a problem.

ASSESSMENT OF EFFLUENT CONCENTRATION AND QUANTITY

The strength and quantity of effluent from a seasonal factory is difficult to assess by analytical means and may require tests carried out over a number of years to arrive at satisfactory figures on which to plan for treatment and disposal. Alternative methods are available. The management must know before it commences operations the intended size of its production of each product and the quantities of raw materials it will require to achieve its targets. In addition the water requirement for each process machine will also be known, together with the requirements for steam raising, product formulation, and factory cleansing. In many cases published figures are available from which to calculate the probable volume of effluent and the quantities of raw materials lost to the drains. In the case of apricots, for example, the percentage loss of fruit on peeling is likely to be of the order of 7·5; of this, 80% is likely to enter the effluent and not be removable by screening or settlement. The dry matter content of the fruit is about 10%, and this is convertible to BOD by a factor of 0·7. Altogether, the probable weight of BOD arising from the peeling operation is

$$0·075 \times 0·8 \times 0·1 \times 0·7 = 0·0042$$

times the weight of fruit peeled, or 0·42 ton per 100 tons of fruit peeled. The volume of water used per ton of fruit will amount to about 500 gal, from which the concentration of the effluent from the peeling process can be calculated as

$$0·0042 \times \frac{2000}{5000} \times 10^6 = 1680 \text{ p.p.m. BOD}$$

(using short tons and Imperial gallons for the purpose of easy calculation). Hence a factory which is to process 100 tons of apricots per day will produce 50 000 gal of effluent at a concentration of 1680 p.p.m. BOD, or a total of 840 lb of BOD, every day it operates its peeling process. Calculations of this nature are essential when considering a proposal for a planned factory which does not yet exist; comparison with published figures is useful to check the order of magnitude derived.

The assessment of the strength of effluents from food factories by the analysis of samples of effluent is a very uncertain business. Analytical figures without flow measurements are virtually useless. The strength may vary tremendously from hour to hour—sometimes within minutes—and in a non-random manner because the strength of the effluent is related to the processes being carried out in the factory and these vary according to a predetermined programme of production. It is therefore necessary to take samples from a balanced flow and to make a simultaneous measurement of the flow. Where a factory is already in production, the better way to assess effluent strength is to strike a waste/production balance. The weight of packed products is deducted from the weight of the ingredients used (including water) and the difference—less any solid waste that can be accounted for—is assumed to have been discharged in the effluent. This approach to the problem emphasises the influence of the exact operation used in promoting waste and effluent production. Numerous analytical figures for the BOD, COD, pH, etc. of food factory effluents have been published, but they vary so widely that they can be of little use for plant design. They may serve, however, as a guide to the order of magnitude that may be expected and to this end a selection of values has been collected into Table 6.1.

The conversion of wastage figures into BOD values can rarely be done with great accuracy because of the variable composition of the materials themselves. Thus the value of 0·453 lb of BOD per lb of potato solids given by Porges and Towne is only a guide; it will vary from season to season, with variety of potato, and even within any one season as a result of changes taking place during storage. Nevertheless, because the same variations in composition must be reflected in any parameter applied to a potato processing effluent it is probably as accurate to assess wastage and use this conversion factor to estimate the BOD load carried in the effluent as to employ any analytical method. It will be appreciated that no flow measurements are required for this assessment.

The factor of 0·7 is applicable to sugars and to a wide variety of other natural substances.

SCREENING
Virtually all food factory effluents are screened as a first step in treatment, and many types of specialised screens are available. It is essential that any screen used shall be self-cleansing, as waste food particles are often gelatinous and tend to adhere to the material of which the screen is made. Microbiological growth is rapid and will contribute to blockage of the screen by growth from the underside if the cleansing is not efficient. Such growth may also intensify any corrosive action of the effluent on the screen material. Of the types in use the brushed screen is the simplest and is suitable for small to moderate flows. Rotating wire mesh screens of drum or disc type are used to remove medium- to large-sized particles; they are cleaned by water jets from the under or reverse side.

TABLE 6.1

Chemical characteristics of some food processing wastes

Nature of process	Location	BOD (mg/litre)	COD (mg/litre)	pH	SS (mg/litre)	TS (mg/litre)
Fruit canning						
Peaches	USA (Calif.)	min. 2 700	3 400			
		max. 4 000	4 200			
Unspecified	UK	min. 140		4·1		
		max. 970		6·8		
Pineapples	Australia	1 200				
Peaches and tomatoes	Argentina	min. 1 800				
		max. 3 300				
Fruit processing						
Peaches and tomatoes	USA (Calif.)	min. 80	170	8·4	60	1 000
		max. 570	1 000	11·3	500	21 000
Citrus	USA (Fla.)	min. 60		6·4	20	270
		max. 400		9·2	110	900
Tomatoes	Argentina	min. 1 400				
		max. 1 900				
Unspecified	Australia	min. 1 500				
		max. 3 000				

			BOD	COD	pH	SS	TS
Canning							
Fish	USA	min.	500	1 300		500	1 000
		max.	4 000	6 000		5 000	20 000
Vegetables	UK	min.	240		4·2		
		max.	1 100		6·8		
Potato processing							
Starch manufacture	Poland	min.	1 600		5·8		
		max.	2 200		6·4		
Lye peeling	USA	min.	2 000	1 200	11·5	2 500	15 000
		max.		2 800	13		
	Holland		2 600	3 500	12·4		
	USA (Idaho)	min.	1 200	3 000	10·6	300	
		max.	5 200	10 000	12·1	6 000	

BOD: Biochemical Oxygen Demand, 5 days.
COD: Chemical Oxygen Demand.
SS: Suspended solids.
TS: Total solids.

Vibratory screens—mechanical or electromagnetic—are suitable for large flows and have the advantage that much smaller meshes can be employed. The NCA has reported on an investigation of single and two-deck circular vibratory screens; a two-deck screen permits the use of many combinations of mesh sizes and is not liable to 'blind'. A 48-mesh, or finer, vibratory screen removed more solids than a 20-mesh rectangular screen and was easier to operate. Bonino, however, has reported on practical experiments at peach and tomato factories which showed that although the removal of suspended solids by a vibrating screen of 80 mesh was of the order of 50%, the BOD was not noticeably reduced. This is because the insoluble solids contribute very little to the BOD. In fact, the author has observed that improving the efficiency of dewatering of the solids on a mesh screen will actually increase the BOD of the effluent as the vibratory action on the screen has a similar effect to squeezing the separated pulp and thus adds more cell contents to the water.

Screening, therefore, is a preliminary treatment which facilitates subsequent handling of the effluent. From the point of view of reduction of polluting strength it may contribute very little. There may indeed be occasions when it would be wise to substitute a comminution process for screening, depending on the degree and nature of further treatment, method of disposal, and climate.

REMOVAL OF SOLIDS

Sedimentation or removal of suspended solids by settlement does often contribute a reduction of BOD up to about 30%. It is often considered necessary if the effluent is to be disposed of without further treatment as the suspended matter might itself cause a nuisance by being deposited in a water-way. Also, if there is biological treatment on conventional lines, the removal of the suspended matter before application of the waste water to biological filters is definitely required. Whether it is essential as a preliminary to treatment by the more modern types of activated-sludge processes is another matter, but since sedimentation tanks act also as balancing tanks, and balancing of the flow is usually necessary, the point is perhaps an academic one.

Sedimentation tanks for food factories are usually of conventional continuous-flow design, having mechanism for the mechanical removal of sludge. It is important that the retention time should not be long enough for fermentation to take place as this will cause rising of the sludge and defeat the object of the process. A problem of this nature may arise at weekends; where there is secondary treatment of the effluent, provision for the return of the purified effluent to the inlet of the sedimentation tank is useful. By recirculating effluent the necessity for the storage of stagnant untreated waste for long periods is avoided. This device also helps to balance the strength of the flow when production is resumed and prevents the application of shock loads to the secondary treatment plant.

A device which combines screening with sedimentation is the use of wedge-wire beds. These effectively remove solids, but have the disadvantage of producing large volumes of watery sludge from many products. They can also be used to increase the concentration of primary and secondary sludges in climates where air-drying is not a practical proposition.

Flotation removal of solids has found applications to the treatment of wastes from potato processing, fruit canneries, and meat products factories. In this process the whole waste water is super-saturated with air and then allowed to flow slowly through a long tank. The air is usually introduced by subjecting a portion of the flow to air under pressure, mixing this highly super-saturated portion with the main flow ahead of the inlet to the separation tank. Release of myriads of minute air bubbles takes place and as they rise to the surface the bubbles raise the suspended matter with them. The solids, including grease and fat if present, form a thick mat on the surface of the tank and this is removed by mechanical surface scrapers. The efficiency of this system depends to a large extent on the nature of the solids in the effluent. If these are heavy—for example, if there is a high proportion of soil—their flotation is more difficult. Applications to the treatment of wastes containing a proportion of oil and fat are self-evident and solids of the nature of hair and skin as well as vegetable tissues with a high water content seem well suited to flotation removal.

Solids of higher density which are not suited to separation by flotation may be removed by centrifugal machines. Hydrocyclones have been applied successfully to effluents from potato processing (usually cold-peeling processes) and concentrate the solids into a thick slurry which can then be further concentrated and dewatered in a centrifugal machine from which the solid material emerges as an apparently dry powder. A physical process of this nature gives a waste solid which is comparatively small in bulk, and in weight, and is much more easy to handle and dispose of than a liquid or semi-liquid sludge.

Chemical aids to precipitation and flocculation have been used from time to time. Usually their success has been limited because they have virtually no effect on the dissolved BOD, they are essentially batch processes unsuitable for the treatment of large flows, and they produce big volumes of watery sludge which may prove an embarrassment. Lime, ferrous sulphate, aluminium sulphate, and aluminium chlorohydrate have been reported upon. More recent trials with the organic polyelectrolytes have shown promise for the treatment of fish processing effluents, but here the objective is not complete purification but to fit them for recirculation of the water within the plant. pH adjustment is a necessary preliminary.

BIOLOGICAL FILTERS

Biological purification of food factory effluents is widely practised. Biological filters of conventional type will purify these wastes as they will purify sewage and at approximately the same rate. Some effluents require fortification with nutrients, being deficient in nitrogen and phosphorus: fruit and vegetable wastes are prominent among the examples. For complete success without ponding troubles the settled effluent must be diluted to about 250 mg/litre of BOD before application to the filters. This is achieved by recirculation of the purified filter effluent in the appropriate proportion. Plant of this nature is only applicable when the factory is in year-round operation, as biological filters must be kept active throughout the year. For factories having high peak seasons in addition to year-round steady production, biofiltration plant can still be used for complete purification of the effluent if provision is made for storage of the peak flows in lagoons from which the waste waters can be withdrawn and purified during the low-flow period.

While effective and reliable, conventional stone filters occupy large areas of land and are very costly to construct. Biological filters using plastics media are much smaller in area and generally less costly overall. They can also be used to effect a partial purification—say, removal of half of the BOD—and they do not need to be fed with a liquid diluted to BOD 250. It is in this region of partial purification that plastics media filters have found their major application. Complete purification to a BOD level of 20 mg/litre or less is possible on plastics media, but not usually in one stage. (*See* p. 26.)

ACTIVATED-SLUDGE TREATMENT

Activated-sludge treatment of food factory effluents is also used widely. The installation cost may be lower than that of conventional bio-filters, but the running costs and maintenance are higher. Activated-sludge plants do, however, require less area. Mechanical agitation and compressed air plants both find their applications. The present tendency is towards operation with higher suspended solids contents than was former practice, but an old plant may need modification of its aeration mechanism if this trend is to be followed. (*See* p. 20.) A modification known as the Kehr process is under examination: this runs with an extremely high content of mixed-liquor solids—of the order of 10 000 to 12 000 mg/litre—and in its original form does not run any sludge to waste. The high-intensity biological oxidation can, as would be expected, deal with shock loadings with a minimum of disturbance. If control of the solids content is exercised this system may well be able to accept seasonal loads from food factories superimposed on a normal town sewage load.

OTHER BIOLOGICAL METHODS

Much less sophisticated, but quite effective, biological treatment plant are in use in various parts of the world. They stem from and are improvements upon the old storage lagoon, which observations showed to effect a biological oxidation. The simplest of these are known as aerobic and anaerobic lagoons; the main structural difference between the two is in depth. They are storage areas of earth-walled construction. In some countries simple earth walls are unsuitable owing to the activities of earth-burrowing rodents which must be controlled in some way if continual embarrassments from leakages are to be avoided. Lagoons of less than 3 ft (1 m) in depth are usually regarded as aerobic. A natural biomass develops in the waste and effects oxidation by processes involving the absorption of atmospheric oxygen and the utilisation of sunlight. The activity depends upon temperature, and a reasonably dry climate is an obvious advantage. Anaerobic lagoons are deeper—4 ft (1·3 m) or more—and the activities of anaerobic organisms in the bottom layers break down the waste solids into simpler compounds which can then be further degraded by aerobic processes taking place in the upper layers in the lagoon. Sometimes the aerobic processes are continued in a secondary aerobic (*i.e.* shallow) lagoon. Gases are also produced by digestion in the lower layers of anaerobic lagoons. Parker and Skerry have described this method of treatment. In a practical application an anaerobic lagoon removed 75% of the BOD when loaded at the rate of 400 lb of BOD per acre per day: an aerobic lagoon was effective at the rate of 50 lb per acre per day.

Lagoon treatment is similar to the cold-digestion process in open tanks formerly used in the United Kingdom for the digestion of sewage sludge. It depends for success on a number of important factors.

(1) Admixture of food factory waste with town sewage is advisable as a means of (a) inoculation, (b) supplying nutrient N and P.
(2) Inoculation with digesting sewage sludge is an advantage.
(3) The climate must be favourable—minimum temperature in the water about 11°C.

The addition of chemical nutrients as well as sewage may be necessary, depending on the nature of the waste.

INFLUENCE OF ALGAE IN PURIFICATION
The term 'anaerobic lagoon' is misleading as it implies a deliberate attempt to exclude air from the process, as in anaerobic sludge digestion tanks; whereas access of air is essential and the top layer of water in a lagoon contains dissolved oxygen. In fact, purification in the upper layer depends in no small measure on the growth of green algae—mainly *Chlorella*—which require ample light for their growth and also produce oxygen. In the bottom layer the conditions are truly anaerobic and digestion, particularly of the deposited solids, is essential. So, unlike the Imhoff tank in which the settled solids are deliberately held in a submerged anaerobic digestion compartment, the lagoon is designed to be of sufficient depth to permit the upper layer of water to shield the lower layers from air while at the same time exposing a large surface area to sun and air. It combines both aerobic and anaerobic processes and is successful in suitable climates for the treatment of strong factory effluents heavily charged with organic solids. It requires, of course, a large area of land. A point which is not obvious but is important is that lagooning over large areas leads to considerable loss of volume by evaporation; this has been estimated to be as high as 50% in Australia.

The use of algae in purification has also been explored by Oswald in California. 'Algal ponds' employ an algal suspension instead of activated sludge. The aerobic ponds are shallow (maximum depth 1 ft (0·3 m)) so as to give the maximum practical exposure to daylight. The liquid is circulated by pumps which induce a spiral flow for the purpose of mixing rather than aeration. Settling of the algae is avoided as photosynthesis is impaired thereby. The algae that develop are natural to the area and the dominant species may vary from season to season. They do not form chains of cells, but, according to species, remain either unicellular or form blocks of four. They convert the nitrogen in the waste into algal protein. Phosphate is also, presumably, essential, but the extent of its removal depends on temperature. In addition to the algae, bacterial zoogloea are also present, together with higher organisms and the microbial mass as a whole may have an ash content up to 30% of its dry matter.

Algal ponds employ eutrophication and algal blooms as a method of purification. Continuous removal of the algal growth and its conversion to dry feeding-stuffs for poultry would solve two problems at once, but there are practical difficulties. In the full-scale applications for treating sewage and a mixture of sewage and winery waste the algal growth is settled out and stored in deep anaerobic lagoons in which digestion takes place and the nitrogen and the phosphorus are resolubilised. The contents of these deep lagoons are discharged during the winter months when river flows are high. Figures 6.3 and 6.4 illustrate aerobic and anaerobic ponds respectively.

FIG. 6.3   Algal ponds at Concord, California.

FIG. 6.4   Storage lagoon for algal sludge.

Bright sunlight is apparently not essential for this system to succeed, but it is not likely to be suitable in areas where temperatures are low or rainfall heavy. Occasionally, spontaneous separation of the algae takes place. This is attributed to a change in pH value and is a disadvantage when it occurs in the pond as it interferes with light penetration.

SPRAY-DISPOSAL

In addition to being non-toxic to sewage purification processes, most effluents from food manufacture offer no danger to public health. Consequently, when it is possible to return them directly to the land, this is the obvious thing to do. In many cases, such as the seasonal effluents produced by in-field installations for the washing of vined peas or of soft fruits, this is the *only* thing to do. Spray-disposal on to pasture is the best practical method; ridge and furrow irrigation can be used in suitable climates.

The object of spray-disposal is to maintain aerobic conditions by avoiding the formation of pockets of effluent—which is the exact opposite of irrigation channels. Effluents from the washing of fruits and vegetables, and indeed any effluent that contains sugars, are prone to anaerobic fermentation, with the production of sulphides and their attendant nuisances. Properly controlled spray-irrigation avoids these conditions.

The first element of proper control is efficient screening to avoid mechanical troubles and also to prevent the deposition of solids as a surface mat which permits anaerobic conditions to develop in the soil beneath. The second is a proper systematic rotation of spraying areas and management of the grass. The third is to have a suitable cover crop. Rose and Mercer recommend the following as an almost universally suitable cover crop in the USA; Ladino clover 1 part; alsike clover 4 to 6 parts; Reed canary, 6 to 8 parts, sown at 13–14 lb/ acre. Alternative grasses are: tall fescue (4–6 lb/acre) and Alta fescue (4 lb/acre).

Szebiotko reporting on the spray-disposal of water from potato-processing plants stated that the increase of hay or cultivated plant yield depends on the quantity of waste applied. Under East European conditions 100-mm doses increased hay yield 147%; 200-mm, 200%; 400-mm, 243%; 600-mm, 265%; and 800-mm, 285%. Increase in yield of potatoes varied from 177 to 213%.

Bonino has also reported favourably on spray-irrigation in the Argentine, particularly of vines and maize. He pointed out, however, that the effluents from some food factories would be too highly concentrated in respect of sodium and mineral salts for application undiluted. Where irrigation is normal practice, the circumspect addition even of these effluents is without much risk. The concentration of organic matter is not a disadvantage as it enriches the soil. Drainage from irrigated areas had a very low BOD and almost neutral pH values. He also stressed the necessity for good horticultural practice and even spraying on a systematic basis; given these requirements 40 000 m³/ha per year can be disposed of in this way in Argentina.

SPECIAL PROBLEMS

Effluents with high salt concentrations arise from many processes in the food industry and often present a problem. Brines from pickled vegetables, from olives, citrus peels, and fruits are particular cases. Where the quantity is large, such brines should be recovered, purified, and re-used.

Where the quantity is not sufficient to warrant recovery, they should be disposed of separately by dumping. Brines containing sulphite in addition will cause particular difficulties if discharged into sewers, because of their corrosive properties, and impose a high oxygen demand on treatment plant. Chemical treatment is required for such effluents involving oxidation and neutralisation, both of which processes can be carried out in automatic plant.

SOLID WASTE DISPOSAL

Solid wastes from the food industry consist largely of trimmings— considered unfit for use by virtue of discoloration, disease, or specific unsuitability—floor sweepings and screenings; they are substandard food. There are several notable exceptions such as coffee grounds, the residue from soluble coffee manufacture, sugar-beet pulp, and pineapple husks, but many of these find a use in some ancillary product or industry. Thus sugar-beet and pineapple pulps form the basis for animal feeds; peach pits are converted to activated carbon for use in water purification. The bulk of the wastes, however, find no use and have to be disposed of. Because they are liable to decompose very easily and will attract vermin, simple dumping on land is liable to cause nuisance and complaints, so they are often consigned to a specialist—usually the local authority—for controlled dumping. Much depends, however, on the climate.

In Australia, waste solids from fruit canning are dumped on land to a depth of about six inches and after a day or so can be harrowed to assist drying before being eventually buried by ploughing. This process is successful mainly because the weather in the fruit canning season is hot and dry. Californian canneries dispose of their solid wastes into the Pacific Ocean. The procedure and its efficiency as a means of disposal form the subject of a report by the NCA Research Foundation, Berkeley, California. The solid wastes from several canneries were consolidated, ground, slightly diluted with water and transported by barges to be dumped into the ocean 20 miles out. The dispersal of the wastes was studied as also were the fish and their behaviour and it was concluded that the procedure has no detrimental effects on the water or on the marine life.

Seafoods processing gives rise to large quantities of solid waste. If the fish could be prepared at sea before landing and the reject materials disposed of overboard, there would be comparatively little difficulty. However, there would still remain quantities of crab shells and trimmings for which disposal by land-fill seems to be the only economic solution. Direct dumping into harbours and estuaries is not an acceptable practice as it takes too long for the materials to disperse satisfactorily. Barging into deep water should produce no problems other than cost as the materials will be used by the ocean's scavengers or will break down on the sea bed; this is provided they are not packed in indestructible plastic sacks before being thrown overboard. The fact that fish wastes have an apparent value as fertiliser or as animal feed has not led to any general development of their potential. This is because the industry is scattered and seasonal, two factors which combine to make by-product industries unattractive.

Because of their nature, food wastes can be composted quite successfully. Some wastes with a high water content need to be partially dried before they are composted as otherwise they tend to collapse into an anaerobic gelatinous mass. An initial water content of 60 to 65% is recommended. Reduction of the moisture content cannot be effected economically by heating. It is done by mixing in a dry organic waste such as rice hulls or sawdust. Optimum sizes of compost heaps, frequency of turning, and general management depend very largely on the prevailing climate. Once a successful cycle has been established and dry compost produced, this material itself can be used as dry diluent for the raw waste, although it will be appreciated that in a Western European climate it is unlikely that the finished compost will have a moisture content below 60%. The process is almost exclusively one of waste disposal: there is little or no market for the compost produced.

It is apparent that most food plant wastes could be digested anaerobically with the production of gas. Comparatively few installations of this type exist, probably because sludge digestion on a small scale using concentrated waste of variable composition is not easy to manage in practice.

BIBLIOGRAPHY
Bonino, A. F., 'Effluents from canneries', 6th International Canning Congress, Paris, 1972.
Cannery Waste Treatment: Kehr Activated-Sludge. Federal Water Qual. Admin., US Dept Interior, US Govt Print. Office, Washington, D.C., 1970. Document ref. 12060 EZP 09/70.
Cannery Waste Treatment, Utilization, and Disposal. State of California Water Resources Board, Sacramento, Calif., 1968. Publ. No. 39.
Dickinson, D., 'Effluents from the industrial preparation of food', *Water Pollution Control* (1967), **66** (2), 159–65.
Kefford, J. F., 'Treatment of wastes from Australian canneries', 6th International Canning Congress, Paris, 1972.
Mercer, W. A., 'Investigation of composting as a means for disposal of fruit waste solids', National Canners Assoc., Berkeley, Calif., 1962.

Mercer, W. A., Rose, W. W., and Doyle, E. S., 'Physical and chemical characterization of the fresh water intake, separate in-plant waste streams and composite waste flows originating in a cannery processing peaches and tomatoes', National Canners Assoc., Berkeley, Calif., 1965.

National Canners Association, Berkeley, Calif., *Research Information* Nos. 164 and 166, 1970; Nos. 184 and 185, 1972.

Porges, R., and Towne, W. W., 'Wastes from the potato chip industry', *Sewage Industr. Wastes* (1959), **31** (1), 53.

Proceedings, First National Symposium on Food Processing Wastes. Federal Water Qual. Admin., US Dept Interior, US Govt Print. Office, Washington, D.C., 1970. Document ref. 12060–04/70.

Graham, R. P., 'Dry caustic peeling of vegetables and fruits', pp. 355–8.

Perkins, G., 'A case history in food plant waste water conservation and pre-treatment experience', pp. 377–82.

Smith, T. J., 'Pilot plant experience of USDA–Magnuson dry caustic peeling process', pp. 359–61.

Proceedings, International Symposium on Utilization and Disposal of Potato Wastes, New Brunswick Research and Productivity Council, Fredericton, N.B., Canada, 1966.

Rose, W. W. and Mercer, W. A. 'Treatment and disposal of potato wastes', pp. 147–74.

Szebiotko, K. 'Total utilisation of potatoes, including the disposal of industrial wastes', pp. 30–44.

Proceedings, Second National Symposium on Food Processing Wastes. Co-sponsored by Pacific N.W. Water Laboratory, EPA, and National Canners Assoc., Denver, Colorado, 1971.

Parker, C. D., and Skerry, G. P., 'Cannery waste treatment by lagoons and oxidation ditch', pp. 251–70.

Soderquist, M. R., *et al.*, 'Current practice in seafoods processing waste treatment', Oregon State Univ., Corvallis, Oregon; for Environmental Proection Agency, Water Quality Office. 1970: Project No. 12060 ECF.

Treatment of Citrus Processing Wastes. Environmental Protection Agency, Water Quality Office. US Govt Print. Office, Washington, DC, 1970, ref. 12060–10/70 WPRD 38-01-67.

Chapter 7

TREATMENT AND DISPOSAL OF WASTE WATERS FROM THE DAIRY INDUSTRY

A. B. WHEATLAND

ORIGIN OF POLLUTING WASTES

Waste waters from the dairy industry are capable of polluting chiefly because of the organic matter they contain and they must normally be treated before discharge to surface waters. The cost of treatment, whether at the dairy or after discharge to a local authority's sewer, depends both on the volume of waste and on the amount of organic matter, that is milk solids, present. The first stage in overcoming an effluent treatment problem, therefore, is to consider whether any measures can be taken to reduce the volume and strength of the waste.

Losses of milk or milk products in the waste waters may result from spillage and leaks; from products left in churns, tanks, piping, and plant of all kinds before hosing clean; from processing, for example, losses from evaporators, butter washing, cheese pressing; and occasionally from deliberate wastage of surplus material such as buttermilk or whey.

Although contamination of waste waters from milk reception depots and bottling plants is significant, the major losses occur from plants producing butter and cheese, evaporated milk, sweetened condensed milk, milk powder, and other milk products.

Before considering how to reduce losses in the effluent, these must be estimated and accounted for by a survey of the dairy operations. Measurement of the volume and strength of all the individual discharges, and cross-checking against the mean volume and composition of the mixed effluent shows where the greatest savings can be made.

PREVENTION OF LOSSES

Losses by spillage and leaks can be prevented by 'good housekeeping'. Those from plant and equipment can be reduced by efficient removal of dairy products before the equipment is washed, and by a pre-rinse with a small volume of water. These concentrated rinsings, which must be collected separately, are sometimes evaporated or, alternatively, they may be used for animal feeding. This pre-rinsing technique can also be used to reduce losses in butter washings. Deliberate wastage of surplus material such as whey and skimmed milk may occasionally be unavoidable, but these liquids should never be discharged to drains. If they cannot be evaporated or used for animal feed they should be removed by tanker and discharged where they can do no harm.

Reduction of milk losses in the above ways requires the continual attention of the dairy management. It is clearly no good reducing milk losses to minimise the size of treatment plant required, if, when the plant has been installed, the effort is relaxed, for the plant will then be overloaded.

Irrespective of whether waste waters are discharged to a public sewer or treated on site, it is usually an advantage to reduce their volume as well as their strength. Cooling water should always be excluded from the contaminated effluent. Reductions in volume can often be made by fitting trigger controls to hoses to ensure that they are not left running, and by using sprays rather than jets for rinsing. Evaporator condensate may be sometimes re-used instead of softened water. There is equipment available for automatically detecting contaminated condensate by means of electrical conductivity measurements.

STRENGTH OF WASTES AND STANDARDS REQUIRED

In considering the most suitable processes to adopt for treatment of the waste waters it is necessary first to know the standard of final effluent required. In England and Wales, standards for discharge to surface waters are decided by the River Authorities after consideration of the local circumstances. Usually they require that the effluent should not have a 5-day biochemical oxygen demand greater than 20 mg/litre. Since washings from plants producing butter and cheese commonly have a BOD as high as 1500–3000 mg/litre (milk itself has a BOD of about 100 000 mg/litre) it is clear that a very high degree of purification (of the order of 99%) is needed. Standards for discharge to estuaries may be lower than for discharge to fresh-water streams, but even estuaries have only a limited capacity for dealing with effluents without pollution becoming evident. Where waste waters are discharged to a public sewer for treatment by a local authority, the conditions are designed chiefly to protect the sewer and to enable the water to be treated efficiently at the local sewage works. Traders in England and Wales have a general right to discharge waste waters into a public sewer, but the law provides that they must first obtain permission from the local authority, which may impose conditions and charge for receiving the effluent.

The high standard of treatment normally required for discharge to surface waters in Great Britain, and the fact that a large part of the BOD of milk washings is due to organic matter in solution, chiefly lactose which is not removed by simple physical and chemical methods, make it necessary to employ some form of biological process after preliminary treatment of the waste.

PRELIMINARY TREATMENT

The preliminary treatment required depends on the character of the dairy waste and whether biological filtration or the activated-sludge process is later to be employed. It may include screening, skimming to remove fat, balancing of flow, and settlement. Screening is especially important at cheese factories to remove curd. Balancing of flow may be desirable, especially at factories working a single shift. Where strongly alkaline solutions are used once per day in cleaning plant, it is usually necessary to store them in a separate tank and to discharge them slowly into the waste waters over a working day in order to avoid an excessively high pH value which could interfere with biological treatment.

Settlement of the waste waters is usually essential. It is common practice to provide a hopper-bottomed settlement tank (Fig. 7.1) which can be desludged without first running off the supernatant liquid. A 4-h period of settlement is usually adequate. Prolonged settlement is not necessarily advantageous because septic conditions rapidly develop and difficulty may then be experienced from rising sludge. The reduction in the biochemical oxygen demand of the waste which is brought about by settlement is usually small—of the order of 15–20%.

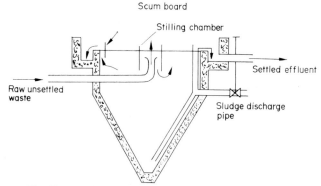

FIG. 7.1   Section of typical upward-flow settlement tank.

After pretreatment and settlement the waste waters still contain some emulsified fat and protein and dissolved organic matter, and must usually be further purified by a biological process such as biological filtration or treatment with activated sludge.

BIOLOGICAL FILTRATION

In this process the waste waters are evenly distributed over the surface of a bed of suitably graded, inert medium such as hard clinker, blast furnace slag, broken rock or even gravel (Fig. 7.2). As the liquid percolates through the bed, bacteria and other micro-organisms grow as a slimy coating on the surface of the medium, feeding on the organic impurity absorbed from the waste. Some of this is oxidised to carbon dioxide, water, and other stable end products, and the remainder is converted to new cell growth. In a mature filter there is a balanced community of bacteria, fungi, protozoa, worms, fly larvae, and other organisms. The protozoa in the film feed on the bacteria, and the worms and the fly larvae feed on the film. Protozoa and metazoa void solids and dislodge pieces of

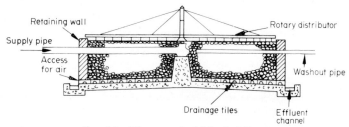

FIG. 7.2   Diagram of a biological filter.

film which are carried away in the effluent and can later be removed, as humus sludge, by settlement. The amount of film in a filter depends amongst other things on the rate of application of organic matter. If the load is too great, or if too strong a liquid is applied, the film grows to such an extent as to seal the spaces between the medium. This causes ponding, the performance of the filter deteriorates, and, because circulation of air through the filter is impeded, the film may decompose anaerobically and unpleasant smells may result.

Many factors affect the performance of a filter, for example, the size, roughness, and uniformity of grading of the medium, the rate, periodicity and method of application of liquid, the depth, the

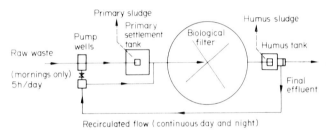

FIG. 7.3   Flow diagram of pseudo alternating double filtration system for a milk reception depot from which wastes are discharged for a short period every day.

degree of ventilation, and the operating temperature.[1] Many types of biological filter exist. They may be used singly, in series, or in series with periodic reversal of the order of the filters.

Experiments by the Water Pollution Research Laboratory from 1933 to 1938, and early experience at dairies, showed that biological filters of the conventional type treating milk washings soon became choked at the surface by heavy accumulations of biological film and fat.[1,2,3,4] It is now generally accepted that single filtration cannot be successfully used for treatment of milk washings unless the waste is first pretreated to remove fat and protein, for example by chemical coagulation,[5] or by lactic fermentation and settlement.[6] The only exception is when the waste is discharged for a short period each day and effluent is recirculated continuously (Fig. 7.3). Such a plant operates in a way essentially similar to that of an alternating double filtration plant, there being a daily change in the use of the filter.

The recommended process for treatment of milk washings to

produce a high-quality effluent is alternating double filtration, with recirculation of settled final effluent to reduce the biochemical oxygen demand of the liquid to be treated to about 300 mg/litre. The waste waters are passed through two filters in series with settlement after each filtration stage (Fig. 7.4). At intervals, usually weekly, the order of the filters but not of the settlement tanks is reversed. The primary filter removes most of the impurity and produces an effluent having a BOD of about 20 mg/litre. When this primary effluent is applied to the secondary filter any surface accumulation

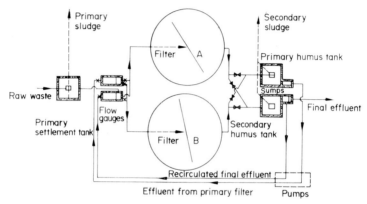

FIG. 7.4   Diagram of an alternating double filtration plant with provision for recirculation of final effluent.

of film which has resulted from its previous use as the primary filter is dispersed and washed out in the effluent. In this way both filters are kept free from troublesome ponding. The quantity of polluting matter which can be removed per cubic yard of medium in the two filters together per day using this method of operation is greater than can be removed from, for example, settled domestic sewage by single filtration. The recommended design loading for an alternating double filtration plant is 0·48 lb BOD/yd³ day (286 g/m³ day)— equivalent to a hydraulic loading of 160 gal/yd³ day (952 litres/m³ day) when the waste has a BOD of 300 mg/litre. Such a plant should be capable of consistently producing final effluent having a BOD of about 10 mg/litre.

The performances of the WPRL experimental filters and of some alternating double filtration plants subsequently constructed are given in Tables 7.1 and 7.2 respectively.

TABLE 7.1

Average results of operation of an experimental alternating double filtration plant
From WPR Technical Paper No. 8[2]

Liquid treated	Rate of application of settled and diluted crude liquid		5-day BOD (mg/litre)				Temperature of final effluent (°C)
	(gal/yd³ day)	(litres/m³ day)	Crude liquid	Settled and diluted crude liquid	Settled effluent from primary filter	Settled effluent from secondary filter	
Milk washings	81	482	820	210	>24	8	7·5
	124	737	1 280	230	14	8	13
	164	976	600	210	7	3	15
	160	952	210	110	11	8	10
	164[a]	976	230	170	15	8	7
	240[b]	1 430	600	300	14	10	17·5
Whey washings	106	630	630	270	29	6	11
	132	785	560	210	13	6	11
	160	952	830	270	28	7	19

[a] Plant operated for only 16·5 h daily.
[b] Period of alternation of filters needed to be reduced from 3 weeks to 10 days to prevent ponding.

TABLE 7.2

Maximum and average conditions of operation of four alternating double filtration plants treating dairy wastes

Plant	A	B	C	D
Volume of waste per day				
Maximum (gal)	160 000	60 000	120 000	90 000
(m³)	725	272	545	408
Average (gal)	130 000	20 000	60 000	40 000
(m³)	590	91	272	181
5-day BOD (mg/litre)				
Maximum	1 500	700	400	1 600
Average	600	300	250	700
Recirculation ratio				
(Raw: recirculated)	1:1	2:3	1:1	1:3
BOD loadings of filters				
Maximum (lb/yd³ day)	0·86	0·48	0·338	0·52
(g/m³ day)	512	286	201	309
Average (lb/yd³ day)	0·36	0·069	0·106	0·102
(g/m³ day)	214	41	63	60
5-day BOD of effluent (mg/litre)				
Maximum	30	10	6	15
Average	8	3	4	4

There are other filtration processes for treatment of dairy waste waters, for example single- or two-stage filtration on a coarse medium with high-rate recirculation of effluent. These processes, however, will not normally produce an effluent suitable for discharge to small fresh-water streams in Great Britain and must be regarded as methods of partial treatment.

USE OF ACTIVATED SLUDGE

In the activated-sludge process the waste waters are mixed and aerated with a flocculent suspension of micro-organisms termed activated sludge. This consists of a variety of bacteria in different stages of development, aggregated together with organic debris, protozoa, and occasionally fungi. The organisms feed on the organic impurity in the waste waters, as in a filter, oxidising some to stable end products such as carbon dioxide, water, and ammonia, and converting some to new bacterial cells. The process is carried out in one or more tanks to provide the necessary period of contact (Fig. 7.5). The mixed liquor then flows to a settlement tank where the activated sludge is separated before being recycled to mix again with the incoming waste. In a properly working plant the overflow from the settlement tanks is clear and is suitable for discharge to a stream. There is a continual growth of activated sludge during the treatment process and when the required concentration in the mixed

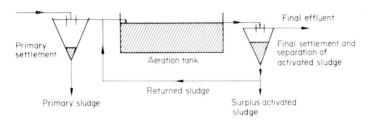

FIG. 7.5   Diagram of activated-sludge process.

liquor has been reached the surplus is removed continuously for disposal. The different types of activated-sludge plant differ principally in the method adopted for aeration of the mixed liquor, the design of tank, and the period of contact of waste with the activated sludge. With filters the most common difficulty is ponding. With the activated-sludge process it is the tendency of the sludge to become bulky. When this occurs sludge does not settle efficiently and may be lost in the final effluent. Also the reduced concentration in the sludge recycled to the aeration tank leads to a progressive decrease in concentration of activated sludge in the mixed liquor and may result in a further deterioration of performance. Bulking of activated sludge is particularly likely to occur in treating dairy effluents.

The activated-sludge process was studied by the Water Pollution Research Laboratory, for treatment of milk washings and washings containing whey, at the same time as the work was carried out on alternating double filtration. It was shown that the process was capable of producing a satisfactory effluent for most of the time, but that it was more liable to upset from shock discharges of strong wastes than was the filtration process. For this reason the filtration process was preferred. More is now known about the activated-sludge process and it is possible that there was insufficient aeration in the early experimental plants to cope with the shock discharges. There have been reports[7] in recent years of the successful treatment of dairy effluents in oxidation ditches (Fig. 3.7). These are essentially activated-sludge plants of simple construction with a long period of aeration. Aeration methods have been in common use for many years in the USA. However, Baumann[8] reports that of 40–50 dairy-waste treatment plants visited only those using trickling filters consistently produced satisfactory effluents. This suggests that the original decision in favour of alternating double filtration for treatment of dairy wastes was justified.

In a survey which covered 384 dairies in Great Britain, responsible for 58% of the country's milk production, it was found that the majority of the dairies (311) discharged their waste to domestic sewers. Of the remainder, 54 operated their own treatment plants of which only one was reported to be of the activated-sludge type. Forty-two of the fifty-four dairies were required to comply with standards of effluent quality laid down by the River Authorities, and in 36 cases a 'Royal Commission' standard (BOD not to exceed 20 mg/litre, suspended solids not to exceed 30 mg/litre) was required. In the other cases more stringent standards in respect of BOD were in force.

OTHER BIOLOGICAL METHODS

Where dairy effluents are discharged to an estuary, it is not normally necessary to purify waste water to the same high standard as for

TABLE 7.3

Conditions of operation and performance of a single-stage high-rate filter (based on data from Morgan and Baumann[9])

Recirculation ratio		
Recirculated flow: raw flow	7·75	3·11
Loading of filter		
lb BOD/yd³ day	1·135	0·67
g BOD/m³ day	675	398
gal/yd³ day	277	133
litres/m³ day	165	79
5-day BOD (mg/litre)		
Raw waste	3 270	1 350
Final effluent	110	50

discharge to fresh-water streams, and single- or two-stage filtration on a coarse medium with high-rate recirculation of effluent is a possible method of treatment. Such treatment may also be useful where it is necessary to reduce the BOD load on an existing alternating double filtration plant or before discharge of waste to a sewer.

TABLE 7.4
Conditions of operation and performance of two plants employing two-stage high-rate filtration with recirculation

Plant and reference	A Morgan and Baumann[9]			B Trebler and Harding[10]	
Recirculation ratio					
Primary filter	1·59	3·28	5·4	—	—
Secondary filter	3·0	5·63	8·9	—	—
BOD loading of two filters together					
(lb/yd^3 day)	1·98	2·1	1·02	1·93	1·0
(g/m^3 day)	1 180	1 250	606	1 150	595
Hydraulic loading					
Primary filter					
(gal/yd^3 day)	1 080	1 080	1 080	—	—
(litres/m^3 day)	6 430	6 430	6 430	—	—
Secondary filter					
(gal/yd^3 day)	1 670	1 670	1 670	—	—
(litres/m^3 day)	9 950	9 950	9 950	—	—
BOD (mg/litre)					
Raw waste	1 190	1 620	1 080	770	740
Influent to primary filter	445	450	210	—	—
Influent to secondary filter	240	190	85	—	—
Final effluent	140	75	30	95	59

The performances of some single- and two-stage high-rate filters reported by Morgan and Baumann,[9] converted from American to British and metric units, are given in Tables 7.3 and 7.4.

Some of the newly developed plastic filter media, although relatively costly, are particularly suitable for a high-rate filter since they have a large void space relative to their effective surface area.

They are increasingly used for reducing the BOD load in dairy

TABLE 7.5

Comparison of the performance of granite and plastic media treating simulated milk waste
(rate of application of waste 1 290 litres/m³ day)

Monthly period	Milk waste (mg/litre)	Loading (kg/m³ day)	BOD				
			Settled effluent				
			Granite medium		Plastic medium		
			(mg/litre)	% removed	(mg/litre)	% removed	
1	927	1·10	161	82	427	54	
2	1 060	1·26	188	82	270	74·5	
3	1 215	1·44	371	82	450	63	
4	1 420	1·68	544	62	495	65	
5	1 930	2·29	787	60	803	58	

waste waters before these are treated in a factory's existing alternating-double filtration plant or are discharged for treatment in admixture with domestic sewage.

COMPARISON OF STONE AND PLASTIC FILTERS

The Water Pollution Research Laboratory has compared the performance of two 7-ft-deep pilot-scale filters, one containing a 2-in granite medium, and the other a fabricated plastic sheet medium.

Reconstituted skimmed milk, prepared from spray-dried powder, was used to simulate a dairy waste and was applied to each filter at a rate equivalent to 200 gal/yd^3 day (1290 litres/m^3 day) without recirculation. The strength of the milk waste was increased gradually during successive monthly periods, from a BOD of about 900 mg/litre to one of nearly 2000 mg/litre. Although the specific surface area of both types of medium was the same (82 m^2/m^3), the granite medium initially produced the better quality effluent (Table 7.5); as the loading increased, however, the efficiency of removal of BOD in the two filters became comparable.

The granite medium accumulated considerable quantities of biological film, and at the higher loadings soon approached the

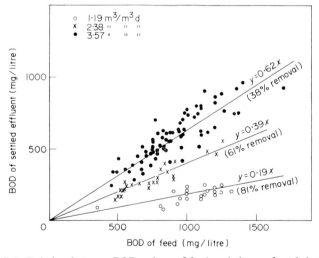

FIG. 7.6 Relation between BOD values of feed and those of settled effluent during treatment of a simulated milk waste at different hydraulic loadings on medium of specific surface area 82 m^2/m^3. (Reproduced by permission of A. M. Bruce and A. G. Boon from Public Works and Municipal Services Congress, 1970.)

condition of ponding which would, in practice, cause serious difficulties. In contrast, the plastic medium, with almost twice the proportion of void space, showed a much smaller accumulation of film, and was never in danger of blocking. This indicates the practical advantages of the plastic medium under the rather severe conditions of loading employed. Although the efficiencies of these filters were high in terms of weight of BOD removed per unit volume, the final effluent was not of the standard normally required for discharge to rivers in England and Wales.

The relation between the BOD of a simulated dairy effluent (reconstituted skimmed milk) and BOD of the effluent from a high-rate filter containing fabricated plastics medium of specific surface area $82 \, m^2/m^3$ for three different hydraulic loadings is shown in Fig. 7.6.

PARTIAL PURIFICATION

Recently the Water Pollution Research Laboratory collaborated with a dairy company in investigating the possibility of partially treating waste waters from the production of powdered and sweetened condensed milk by an aeration process. The need was to achieve a 65% reduction in the BOD of the waste to enable a greater volume of liquid to be treated in admixture with domestic sewage at a nearby sewage works. The process investigated was that known as contact stabilisation, in which the waste is aerated for a short time in contact with activated sludge. The sludge is subsequently removed by settlement and reaerated for a relatively long period before mixing again with the waste (Fig. 3.3).

Initially, experiments were made using a loading of about 155 lb BOD per 1000 ft^3 contact plus aeration capacity per day (2·48 kg/m^3 day), the corresponding volumetric loading being 10 300 gal/1000 ft^3 day (1650 litres/m^3 day), but difficulty was soon experienced with bulking of the activated sludge. A reduction of the BOD load to one-third prevented further deterioration of the settlement characteristics of the sludge, but on increasing the load to the original value serious bulking again occurred. It had been observed in laboratory experiments that there is a range of BOD loadings (lb per day per lb activated-sludge solids) in which bulking of activated sludge fed with domestic sewage is troublesome and that at relatively low loadings and high loadings bulking is not experienced. The load on the experimental plant was therefore increased threefold to 31 300 gal/ 1000 ft^3 day (5020 litres/m^3 day), raising the BOD loading to 470 lb/ 1000 ft^3 day (7550 g/m^3 day). These values corresponded to a contact time of 1·6 h, based on the flow of waste water, and a sludge re-aeration time of 3·1 h. The concentration of activated sludge in the mixed liquor was maintained at about 8000 mg/litre and the sludge loading was thus about 0·57 lb per day per pound of solids in the contact and re-aeration tanks. Under these operating conditions the sludge–volume index (ml/g) decreased to a value in the range 50–100 indicating that the sludge had good settling properties. Over a period of 2 months the BOD of the milk waste was reduced from about 1500 mg/litre to about 500 mg/litre. In a subsequent 4-month period the rate of flow of waste water to the plant was varied daily in a predetermined pattern to simulate the variation in the rate of flow from the factory. Satisfactory results were maintained. The sludge from the re-aeration compartment had a specific resistance to filtration in the range 0·01 × 10^{13} to 5 × 10^{13} m/kg. Aerobic digestion of the sludge tended to increase its specific resistance to filtration and it was thought that this effect would outweigh any advantage to be gained by the slight reduction in the quantity of sludge requiring disposal.

CHEMICAL COAGULATION

Treatment with hydrated aluminium sulphate (about 500 mg/litre) and lime, followed by settlement, will reduce the BOD of milk washings by up to 85%.[2] The clarified liquid can then be further treated by single filtration. At one plant visited by Laboratory staff, 300 mg aluminium sulphate was used per litre of waste. After settlement, the liquid was treated at a loading of 0·24 lb BOD/yd^3 day (143 g/m^3 day). The effluent had a BOD of 14 mg/litre.[5]

Treatment with a coagulant is relatively costly and gives rise to a large volume of watery sludge which, because of its aluminium content, is unsuitable for disposal on cultivated land.

SCARLETT'S PROCESS

In the early work on the treatment of dairy effluent it was observed that if washings containing 1% milk (BOD about 1000 mg/litre) were allowed to ferment at a temperature of 30°C for 1 day they could subsequently be treated by single filtration at a rate of 100 gal/ yd^3 day (0·6 m^3/m^3 day) without trouble from ponding. Washings which had not been allowed to ferment brought about choking of the filter under similar conditions.

This idea was developed by Scarlett[6] who suggested aerating the milk washings at a temperature of about 30°C to encourage lactic fermentation and coagulation and precipitation of the fat and protein (Fig. 7.7). A number of full-scale plants of this type have

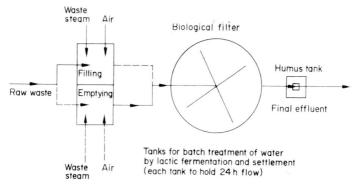

FIG. 7.7   Flow diagram of Scarlett's process involving preliminary coagulation by aerobic lactic fermentation and settlement.

been built.[11] A noteworthy feature of these plants is that dilution is unnecessary. However, the process is probably relatively costly since it involves the provision of two tanks each large enough to hold 24 hours' flow of waste water, as well as a biological filter and settlement tank.

SPRAY DISPOSAL ON GRASSLAND

Disposal of dairy effluent by spraying on grassland is practised extensively in the United States,[12] New Zealand,[13] and Holland,[14] and is used to a limited extent in Great Britain. The method is not suitable for use during prolonged periods of frosty or wet weather, and in the British Isles its chief value would be to cope with the extra volumes of waste during the flush season in late spring and early summer. Pretreatment of the waste by screening and balancing of flow is normally required.

A site for spray irrigation should be flat or only slightly sloping, preferably to the leeward of the factory and any dwellings. Care must be taken to avoid trouble from wind-borne droplets. Low pressure sprays minimise this. The soil should preferably be sandy and well drained. A cover crop such as grass is required.

McDowell reports that it is usually convenient to graze or harvest the grass ahead of spraying and then to allow 10 days for the grass to recover before spraying is begun. A 14-day rest period is allowed after spraying; the grass is thus irrigated once every 25 days. The amount which can safely be applied depends on the weather and condition of the soil. It will usually be equivalent to 0·15–1·0 inch rainfall (0·38–2·54 cm), that is 3400–22 600 gal/acre (0·0038–0·0254 m^3/m^2). For details of procedure and precautions, the publication by McDowell and Thomas[13] is very helpful.

DISCHARGE TO PUBLIC SEWER
Where sewers are available it is usually best for arrangements to be made for dairy effluent to be treated in admixture with sewage at the local works. It is important that the works should be capable of dealing with the maximum peak season load from the dairy and that the discharge should be controlled so as to avoid shock loads reaching the sewage works. By-products such as whey, skimmed milk, etc. should never be discharged to the sewer.

If the dairy effluent represents a large proportion of the total flow of sewage, the sewage may become septic in the sewers and in the primary settlement tanks, and cause odour nuisance and sulphide attack of concrete. In such a case it may be desirable to convey the dairy effluent separately to the works and settle it before it is mixed with the sewage.

SLUDGE DISPOSAL
It is usual for primary, humus, and surplus activated sludges from plants treating dairy effluents to be dewatered on drying-beds before final disposal on land.

The sludges can be dewatered on filter presses but it is normally necessary first to 'condition' the sludge with chemical coagulant. Aluminium chlorohydrate added in a concentration of about 2% (as Al_2O_3) on dry weight of solids was effective with crude primary sludge and with surplus activated sludge in some of the Laboratory's experiments.

ACKNOWLEDGEMENT
Crown copyright. Reproduced by permission of the Controller of Her Majesty's Stationery Office.

REFERENCES

1. Water Pollution Research Laboratory. Notes on Water Pollution No. 5, June 1959; No. 14, September 1961.
2. Department of Scientific and Industrial Research. Water Pollution Research Technical Paper No. 8. HMSO, London, 1941.
3. Jenkins, S. H., *J. Proc. Inst. Sew. Purif.* (1937), Pt. 1, 206.
4. Jenkins, S. H., *J. Proc. Inst. Sew. Purif.* (1937), Pt. 2, 13.
5. Southgate, B. A., *Dairy Sci. Abstr.* (1954), **16,** 427–42.
6. Scarlett, C. A., *Analyst, Lond.* (1939), **64,** 252–60.
7. Schaafsma, J. A. A., *Zoetermeer Algemeen Zuivelblad* (1957), **50,** 306–9, 330–2.
8. Baumann, E. R., In *Wastes Treatment* (Ed. P. C. G. Isaac). Pergamon Press, Oxford, 1960.
9. Morgan, P. E., and Baumann, E. R., *J. sanit. Engng Div. Am. Soc. civ. Engrs* (1957), **83,** No. SA4, Paper No. 1336.
10. Trebler, H. A., and Harding, H. G., *Proc. 4th Industr. Waste Conf.,* 1948. Purdue Univ. Engng Extn Series No. 68, 67–79.
11. Crossley, E. L., *Report to International Dairy Federation,* 1953.
12. Lawton, G. W., Breska, G., Engelbert, L., Rohlich, G. A., and Porges, N., *Sew. Industr. Wastes* (1959), **31,** 923.
13. McDowell, F. H., and Thomas, R. H., N.Z. Pollution Advisory Council, Publ. No. 8, 1961.
14. Baars, C., Graaf, A. W. D., and Keuning, J. A., Rijks Zuivel-Agrarische Afvalwater dienst, Arnhem, Publ. No. 14, 1960.

Chapter 8

THE MANAGEMENT OF FARM WASTES
DENIS DICKINSON

Although all races engage in the rearing and management of animals for food in all regions of the world, methods, practices, and customs vary so greatly that there is no possibility of defining a standard or even typical farm or method of farming, which would apply to anything more than a small geographical area with a uniform climate. Consequently, it must first be made clear that the information in this chapter applies primarily to British practice under British conditions of climate and geography and that the technical data derive almost exclusively from British sources. The recommendations made will apply to other temperate zones where practices are similar, but differences due to climate, feeding methods, availability of water, breed of animals, population densities and so on must always be anticipated.

Technical data are comparatively scarce, for although river inspectors and others with a special interest have been aware of the occurrence of water pollution by farm wastes for a long time, until 1961 such events were accepted almost as natural hazards in much the same way as flood or drought. The Public Health Act of 1961, which applied only to England and Wales, brought farm effluents into the category of trade effluent, while the Rivers (Prevention of Pollution) Act of the same year gave River Authorities in England and Wales certain powers to control all farm effluents discharging into rivers. These changes in the law followed radical changes in agricultural practice which had begun during the post-war period, and they also coincided more or less with a change in the public attitude towards water pollution in general. There was a certain unease at the apparently indiscriminate use of agricultural chemicals and the dumping of wastes in general.

How little was known about waste waters from farms at that time is clearly illustrated by the publication 'Waste Waters from Farms: Notes on Water Pollution No. 17' issued in June 1962 by the Water Pollution Research Laboratory of the Department of Scientific and Industrial Research. This leaflet of less than four pages was published in response to 'requests for information on the polluting character and methods of disposal of farm effluents . . .' and its purpose was '. . . to set out briefly what is known.' This was clearly inadequate and the issue of the leaflet at that time was important because it showed that research and investigation were urgently necessary if the new laws were to be implemented. A second Note, No. 24, issued in March 1964, gave further information derived from new investigations mounted by the Water Pollution Research Laboratory. This made it clear that the Laboratory had responded to its own message, but that other bodies—particularly those primarily concerned with agriculture—would need to join in if the problems revealed were to be solved within a reasonable time. Subsequently, upwards of 30 original papers contributing new information have been published in the British technical press and there have been a number of seminars and symposia in addition. The following pages attempt to summarise this information.

SILAGE LIQUOR

Water pollution by silage stack drainage can be spectacular owing to the very high strength of the liquor. It is acid, has a high immediate oxygen demand, and a 5-day BOD of about 50 000 mg/litre—at which level accurate assessment of the BOD becomes not only impossible, but pointless. There have been instances of sewage purification works being put out of action by the shock loads resulting from the discharge of silage liquor into sewers; of water supply sources temporarily discontinued through contamination by silage stack drainage; and of massive fish kills resulting from its discharge to rivers. Such events have helped to focus attention on farm effluents in general. Silage 'juice' or 'liquor' is the drainage from a stack of green vegetable matter—grass, clover, and pea-haulm are the most usual—and it consists of fermented sap and adventitious moisture expressed from the material by its own weight or external pressure. As the purpose of ensilage is the preservation of the vegetable matter as feeding-stuff, the liquor constitutes a direct loss of nutrients and is avoided as far as possible. The dry matter content of the greenstuff at the time of ensiling is the factor which controls the quantity of liquor which drains from the stack. Gibbons (1968) quoted figures from the Ministry of Agriculture, Fisheries, and Food to show that if the dry matter content of the ensiled material were 25% or more, the volume of liquor draining from a covered silo would be negligible. Previous figures of Jones and Murdoch (1954) were less optimistic; they referred to grass and reported that an increase in the dry matter content of the material on ensilage from 17–20% to over 26% was accompanied by a reduction in the volume of stack drainage from over 40 to about 2 gal/ton of material. This desirable increase in the dry matter content is achieved by wilting the material in the field before stacking, a procedure which is highly recommended. Wilting is sometimes necessary to promote desirable fermentation in the silo; if the sugar content of the sap is too low the desired bacterial action does not take place, but the position is most easily rectified by wilting the grass. An alternative procedure is to add sugar in the form of diluted molasses, but this increases the volume of stack drainage. It may be anticipated that under adverse weather conditions wilting may not be possible in the field and the addition of molasses may be necessary with consequent difficulties with silage liquor. Little (1966) referred to another objection to wilting: the system involves a further handling stage and therefore requires more labour. There is also a maximum dry matter content above which satisfactory silage is not always obtained; this is around 27%.

It is apparent that the farmer, whose prime objective is the production of a satisfactory silage, may not always be able to take the steps which the river inspector believes to be necessary; but they should not disagree over the matter of protecting the stack from the rain. If rain is allowed free access to the ensiled material the stack drainage must increase with consequent loss of nutritive value and probable interference with the course of the fermentation.

By using a wilting procedure and covered silos, the volume of stack drainage can be reduced under favourable weather conditions to less than three gal/ton of material ensiled. During a wet summer the volume may be expected to increase up to 10 or 20 times this amount. The liquor is not easy to store as its composition favours the growth of moulds. If it has access to drains or soakaways these are liable to become blocked by mould growths. The liquor has a certain value as a fertiliser, containing useful concentrations of nitrogen, phosphorus, and potassium, but it cannot be distributed on land without dilution as it is acid and causes scorching. Even so the best method for disposal of silage liquor seems to be spray-disposal on to land after suitable dilution. On a farm where there are other effluents to be disposed of in the same way, the solution is to add the silage liquor in small quantities as it is produced to the bulk of the effluent and to spray the mixture. If, unusually, there is no other effluent, the liquor must be diluted with water before spraying. Every precation must be taken to ensure that it does not go either directly or indirectly through land drains into any surface water.

Some Local Authorities are prepared to accept silage liquor for purification in their sewage works. It may be transported there either *via* the sewers or by tanker. Because of its strength the Authority will insist on a strictly controlled rate of discharge and may also limit the period of the discharge so as to avoid overloading the purification works.

Whatever method of disposal is adopted, it is obvious that the liquor must first be collected from the silo. Efficient drainage into a separate collection tank is a pre-requisite. The tank need not be elaborate—Little (1966) described the use of concrete pipes sunk into the ground—but it must be watertight. As more silage liquor may be anticipated during wet summers it might be wise to provide for the collection of rainwater from farm buildings into an adjacent tank; this could then be used for dilution of the liquor before spraying.

ANIMAL WASTES
Cattle
Beef and dairy cattle excrete almost 10% of their own weight each day in dung and urine. When the animals are ranging freely over pasture, these excretions are returned directly to the land and under normal conditions of good husbandry no particular problem either of waste disposal or of water pollution arises. Problems do arise when the animals are massed together in farm buildings or enclosures. The nature of the problem there depends to a large extent on the system of management adopted. If the cattle are housed in yards with sufficient straw to absorb a proportion of the water content of their dung and urine—which is normal practice when wintering beef

cattle—their waste can be handled as solid matter, stacked and spread in the traditional manner. Obviously, rain water must be excluded as far as possible from both the yard, or court, and dung heap as it will nullify the absorptive drying effect of the straw and also leach out a proportion of the manurial value of the heap. If the amount of adventitious water derived either from rain or from yard washings is sufficient to cause liquid to seep from the dung heap, this drainage will itself constitute a problem.

The quantity of water which collects on an acre of roofing or concrete yard is approximately 22 000 gal/in of rainfall and as this is essentially clean water provision for its separate collection should be made; it can be stored in a reservoir of some kind for use as supplementary water supply or it can be led away by ditch to enter a stream at a convenient point. Developments in the mechanical handling of manure have ensured that even at very large units where the animals are housed on straw, no effluent disposal problems have been experienced. Disposal of the manure, however, requires a sufficient area of land for its utilisation.

With dairy cattle, the position is different. The use of straw as an absorbent has declined and the most usual system is now to collect the droppings as a slurry. The combined water content of the urine and dung voided by cows is 85–90% so that the mixture is a liquid which will flow. It can be collected in channels or caused to flow through slats into underground collecting tanks. To dispose of this slurry by spraying on to land it is necessary to dilute it with about an equal volume of water so that there is room in the collection system to accommodate about 3 or 4 gal of water/cow/day; this volume can therefore be used for the various washing operations necessary in managing dairy cattle without adding unnecessarily to the waste disposal problem. It follows that the collection system must be of adequate size.

Opinions differ as to whether the wash-down water should be collected with the slurry or kept entirely separate. If it is combined, then the volume/cow/day under winter conditions will be at least 12 gal and as this may need to be stored for as long as 13 weeks, the storage capacity required per animal will be 1092 gal. The size of the herd and local conditions are both important factors here. Separate collection and disposal of wash-down water introduces other problems. It has been stated that such water may be disposed of by passing *via* a channel with simple filtration into a ditch or soakaway within the boundary of the farm, but this appears optimistic. Wash-down waters are often quite strong, and also the wrong things have a way of getting into separate collection systems (other than roof drainage systems which can be sealed at ground level) so that washings may well require a complete treatment plant.

It is supposed that the farmer will wish to make the best use of the manure available. It has a considerable value as a fertiliser in terms of NPK whilst its value as a soil conditioner and growth promoter is universally recognised but not quantifiable. For proper utilisation, as opposed to mere disposal, storage capacity is essential. A land area of about 0·5 acre/cow is required, which is not very different from the area required by free-ranging cattle. Further information on the rates and methods of application, effects of storage and similar matters is given in the section on Slurry Management, p. 137.

Pigs

The pig is no longer the poor relation of the farmyard, but is now reared almost exclusively indoors in specially designed rearing stations. The pig population of Britain increased almost fourfold during the 20 years from 1948, now standing at about 8 millions. The intensive system of rearing means that the majority of these animals are kept in houses over a slatted floor beneath which is a dung channel; this allows their droppings to be removed from below by either mechanical or hydraulic means. For ready hydraulic removal with a minimum of water, the channel should have a slope of about 1 in 50. The quantity of waste produced depends on several factors which make it difficult to state an average; a sow may weigh three times as much as a pig, yet produce about the same quantity of waste. This is about 2 gal/day on dry feeds, but will increase to 3 gal if the animal is fed on swill. The one figure on which there is a considerable measure of agreement is the 'population equivalent value'; one pig produces from 2·5–3 times the BOD load of one person, i.e. about 0·40 lb BOD/day. It follows that the pollution potential of the pig population of Britain is almost half as great as that of the entire human population.

Poultry

The intensive rearing of poultry in batteries is a long-established practice which has been extended of recent years to include turkeys and ducks. It is necessary to distinguish between birds raised for egg production and table birds. The greatest number of hens are laying hens, usually cage birds, and they produce from 1–2 tons of manure/ 1000 birds/week. There is often an unavoidable loss of drinking water into the waste and Riley (1968) gave figures of 1 ton/week from 1000 layers, containing 75% moisture, weighing 65 lb/ft^3 and measuring about 1 m^3. With this composition the manure is neither liquid nor solid, but plastic, and consequently difficult to handle. For pumping as slurry it is necessary to add an equal weight of water.

Wastes from deep litter or broiler birds are much drier, spreadable on land as solids, and generally present no problem. Turkey manure is also drier—moisture content about 55%—and presents no particular problem. Ducks, however, must have water and the waste from duck rearing is a slurry containing 95% of water, which can be pumped away for disposal.

Poultry wastes do not appear to be suitable for biological purification and no reports have been seen of their treatment with domestic sewage in municipal plant. BOD values are therefore only of interest in regard to possible water pollution and as the manures are more solid than liquid, the BOD will vary directly with moisture content. The BOD load production by 1000 birds is of the order of 20 lb/day.

Much attention has been given to the drying of poultry manure for general use in horticulture and as a base for fortification with chemical fertilisers. The dried material contains respectively about 5, 3 and 2% of N, P, and K. Expressed in other terms by another authority, the value of the manure produced by 1000 birds is currently £100/year. As its moisture content is already comparatively low, poultry manure can be dried to stabilisation point at much less expense than any other animal manure.

At the other extreme, recent experiments have been carried out using a channel or lagoon of water beneath the battery house to catch and digest the droppings. The gases produced by digestion of the organic matter must be removed by the ventilation system, and the slurry is eventually pumped away for spraying. A report states that such experimental lagoons have been left for 400 days without emptying (Jones, J. L., 1973).

Probably the greatest difficulty associated with poultry wastes is their offensive smell. This increases quite rapidly with storage and as storage is essential during winter the problem is widespread. No entirely effective solution has yet been recommended.

MINK FARMS

Mink farming is practised extensively in the British Isles. The animals are housed in wire cages raised 2–3 ft from the ground and partially protected from the weather. The number of animals ranges from a few hundreds to 20 000 or more on a large farm. The amount of waste produced by the animals is not very great but it is supplemented by waste food and drinking water, creating a problem. The cages are usually in long rows covered by a roof and as the structure may cover a large area the quantity of water collecting from the roofs during rain is quite considerable. If this is not collected but is allowed to fall on to the ground and soak away beneath the cages it becomes heavily polluted before it enters the land drains. Provision of a proper drainage system to collect the rain water and the drinking water spillage and other waste from the cages is a necessity on a large farm. Separate collection of the rain water may not be an advantage as the waste from the cages is extremely strong and has a high solids content; it requires dilution to make it flow and to make it treatable.

Mink are fed on very special diets containing fish or fish meal, tomato purée, cereals, and a range of supplements. Food preparation is therefore carried out in a special building from which washdown and other wastes accrue as from a food factory. The effluent from such buildings can be very strong indeed. The mink are fed in their cages individually; they are also watered individually and the amount of waste, although small per animal, can be quite large in total.

The drainage from a mink farm may have to receive complete purification before discharge into a stream. Land disposal is often not practicable because of the nature of the terrain or because a sufficient area is not available. The collected wastes may have to be diluted in order to convey them to a site for treatment. The suspended solids content and the fat or oil content are both higher than is usual in an organic waste water. Preliminary screening is advisable, followed immediately by solids removal. The solids tend both to settle and to float so that a conventional settling tank can become covered with a thick mat of floating matter. If the size of the enterprise warrants the expenditure, the waste could best be treated by flotation of the solids; otherwise, duplicate 'solids removal' tanks are needed so that a tank can be emptied completely to collect its settled and floating sludge on to underdrained drying-beds. Once separated as sludge, the waste solids drain and dry on beds quite satisfactorily and can then be disposed of on land as manure. The liquid waste can be purified in percolating filters, after dilution with recirculated filter effluent. This recirculation is essential as there is not usually sufficient volume of the waste to effect efficient distribution over a filter. A recirculation ratio of at least 5:1 is recommended. Settlement after filtration is necessary in the usual way.

SLURRY MANAGEMENT

Farm slurries are disposed of by spraying on to land, thus continuing the ancient practice of returning waste to the land to maintain fertility. The main difference between the older system of stacking, or composting, followed by spreading of the material as solid and the present system of storage of the slurry in tanks followed by its distribution by spraying lies in the changes that can take place during the period of storage. The older system involved much man-handling and the creation of a dung heap which would be marginally aerobic. Microbiological decomposition took place with the genera-tion of much heat and temperatures were reached which were suffi-ciently high to destroy weed seeds and probably certain pathogenic organisms. On the debit side, the heat was probably sufficient to administer the heat shock necessary for certain microbiological spores to germinate. During slurry storage the conditions are anaerobic, some digestion of organic matter undoubtedly occurs, but there is little elevation of the temperature. Many have expressed doubts about the wisdom of spreading fresh slurry and storage intervals of 12 days to 12 weeks have been recommended, during which period it is expected, or hoped, that the numbers of patho-genic organisms will decline considerably. This may be so, but the fact remains that there are no figures to confirm or deny this opinion, that no one can say what is the safe level of bacterial concentration that may be sprayed—10^6 or 10^{12}/ml—if indeed there is such a thing as a safe level of concentration in this connection, and there is no information whatsoever regarding the transmission of viruses in this way.

It has been pointed out that if a farm or factory farm is fitted with a continuous slurry collection, storage and disposal system, an outbreak of disease among the animals can lead to contamination of the whole system with the vector responsible *before* the disease is detected. The only advice offered to meet such a contingency was to store the slurry for as long as possible, to spray it on areas not being grazed, and to allow the maximum possible period for weathering between spraying and grazing or between spraying and cropping for fodder. This situation is not by any means satisfactory.

By contrast there is quite a lot of information about the use of slurry as fertiliser. The papers by McAllister (1970) and by Wheatland and Borne (1970) in particular contain useful figures. However, it is apparent that because of differences in management methods it would be misleading to quote average values for fertiliser constituents and each farmer will need to solve his own problem with the help of the National Agricultural Advisory Service. In fact, the problem which seems to be developing from the use of slurries is one of *over*-fertilisation and in particular the return of excessive quantities of nitrate. Related to this may be land damage arising from the physical effects of spraying and/or the machinery involved, as well as that due to the chemical constituents of the slurry, such as nitrate and phosphate. The situation is likely to become acute when an area of land is used for the disposal of the waste from more animals than it could naturally support. At this stage the animals are being fed on materials brought in from elsewhere and the imported feeding-stuff contributes to the quantities of nitrogen, phosphorus, and potassium returned to the land.

How much and how often are not questions to which any simple answer is possible. It is not even to be expected that the same quantities may be applied to the same land in successive years. The guidance of the agricultural scientist and the judgement of the farmer must prevail over the wishes of the anti-pollution expert.

BIOLOGICAL TREATMENT

Animal wastes and the general non-chemical drainings from farms are obviously of an organic character amenable to biodegradation. It is now established, however, that farm sewage is more difficult to purify than domestic sewage. An indication of this is contained in the chemical analysis; the ratio of permanganate value to BOD is much higher in the case of farm effluents. Furthermore, in a biologically purified farm effluent this ratio is even greater, indicating the presence of organic substances which are resistant to biological degradation. Some of these substances are highly coloured so that treated farm effluent or a mixture of sewage and farm effluent is brown in colour; it will also have a comparatively high permanganate value even though its BOD value is satisfactory. Local Authorities that accept farm effluents into their sewers for subsequent purification and make a charge for this service have used a 'treatability factor' to allow for the acknowledged difficulty of treatment in comparison with domestic sewage. This factor has sometimes been based on the BOD/PV ratio, but as this ratio is variable it is not regarded as satisfactory for the purpose of assessment of charge. It has been suggested that the actual performance of biological processes should be used as the basis of charge, but this is not easy to assess. For percolating filters the dosage rate for farm effluent has been reported to be about one half of the rate for domestic sewage to achieve an effluent of the same quality (BOD 20 mg/litre) so that a cost treatability factor of 2 would be appropriate for this type of plant. Difficulties in primary treatment (sludge settlement) and in sludge digestion due to farm wastes have also been reported, but no figures are available on which to base a treatability factor for these processes. It will in any case depend upon the methods of stock management adopted by the farmer, on the ratios of the various animals kept, and on the methods used for solids collection and removal on the farm so that no universally applicable treatability factor is to be expected. The Local Authorities' approach to the matter of charges and the basis on which these are made are explained in a paper by Simpson and Hibberd in 'Farm Wastes; A Symposium'.

BIOLOGICAL FILTRATION OF FARM EFFLUENTS
Although it may be anticipated that conventional biological filtration with recirculation of the effluent could be used with confidence to purify farm wastes, there are few reports of the method being used except at experimental farms and research stations. The reason is the cost and size of the plant that would be required. Some $2 m^3$ of filtration medium would be needed per cow, probably 1 to $1 \cdot 3 m^3$ per pig, with a recirculation ratio of about 3:1 and settling tanks and sludge disposal facilities in proportion. Add to this the generous contingency factor that would be necessary to ensure that the plant would function efficiently without constant attention and the installation becomes both too costly and too big for normal circumstances.

The use of high-rate biological filters of both wood and plastics construction has been investigated experimentally and reported by

Hepherd and Charlock (1971). Both the objective and the methods used were unconventional as the liquid fed to the filters was mainly piggery slurry with a dry matter content averaging 3 %. It is perhaps rather surprising that very high proportions of the dissolved BOD and COD were removed under the conditions used and the facts reported lend further support to the contention that high-rate filtration through towers in this way is more similar to an activated-sludge process than to bio-filtration proper. What is not surprising is that mechanical blockages caused practical difficulties and although the process shows promise there is much more development work to be done before it can be recommended for adoption on farms. The most likely application for such a plant would be at a pig rearing station where sufficient land was not available for disposal of the slurry, but there would still be the problem of disposal of wet sludges, even though concentrated to some extent and deodorised by the process.

ACTIVATED-SLUDGE TREATMENT

The purification of effluents from pig farms has been investigated more particularly in Holland and papers on their treatment in oxidation ditches have been published by Scheltinga (1966 and 1969); Pontin and Baxter (1968) have reported on their work in Britain. Sidwick (1972) has described a full scale installation in Holland which treats the effluent from a pig-rearing station housing 11 000 pigs. It is obvious that a farm with this number of animals would require too great an area of land to dispose of the effluent directly as slurry; it has been estimated that the waste from seven animals requires one acre of land for its disposal as slurry. The system applied is a modification of the extended aeration system as it has been found that the attainment of an effluent of BOD 20 mg/litre and oxidation of surplus sludge would require far too long an aeration period. Instead, a combination of lagooning and aeration in a conventional oxidation ditch with a nominal detention period of 10 days produces a highly coloured effluent with a BOD of about 80 mg/litre which is satisfactory for disposal on to land without causing odour problems or water pollution. Pontin and Baxter have described an experimental installation catering for 600 pigs in which the usual under-floor dung channels were modified to form one continuous channel; this was fitted with aeration rotors and constituted a combined collection system and oxidation ditch. A secondary oxidation ditch, constructed outside the housing, provided further treatment. This system appears to be reasonably economic, but the effluent produced was only partially treated, having a BOD of about 600 mg/litre. The system avoids some of the disadvantages reported elsewhere. There were no difficulties with foaming; toxic gases which emanate from anaerobic collection systems were avoided; and the risk of transmission of pathogenic organisms was reduced considerably.

ANAEROBIC DIGESTION

It is common practice to store slurries for quite long periods before spraying and during such storage digestion takes place resulting in loss of nitrogen as ammonia and the production of carbon dioxide, methane, and hydrogen sulphide.

Casualties due to the inhalation of these gases have occurred among the animals and precautions must be taken to prevent the gas concentrations rising to unsafe levels. Disturbance of the stored slurry releases gases. It has also been reported that in covered storage tanks the concentration of methane in the air space has reached the danger level of 5–6% at which there is risk of explosion. There is plenty of incidental evidence that farm wastes are digestible with the production of utilisable gas. Pontin and Baxter (1968) referred to a prototype plant exhibited in 1963 by Wright Rain Ltd which was intended to produce 10 000 ft³ of methane/day from the waste from a 1000-pig unit digesting anaerobically at 35°C. At this rate there would be a considerable surplus of useful gas. The process has been used on a small scale from time to time—Waldemeyer (*Wat. Poll. Cont.* 1970, **69**, 178) refers to one instance and it is notable that in discussions of papers presented to the Institute of Water Pollution Control experienced members of the Institute have frequently urged the adoption of alkaline anaerobic digestion for animal wastes.

CATTLE MARKET WASTES

Cattle markets—known as 'Marts' in some parts of the British Isles —vary in size and frequency of occurrence from small town fairs, usually held in the streets of small country towns at infrequent intervals, to the regular ordered assemblies in cattle centres.

Where no special market premises exist, the beasts are marshalled in the streets and their droppings are distributed indiscriminately.

On wet days, these wash continuously into the surface water sewers; on dry days the market is usually followed by a clean-up which involves hosing down the roads and pavements. Pollution of the local receiving water is a virtual certainty in these circumstances and there is little that can be done about it. In course of time, however, these small town fairs will almost certainly disappear.

Organised markets are usually constructed over sewered areas and may be partially or wholly covered against rain. A modern market will have separate drainage for the cattle pens and for the roof waters, thus permitting a measure of control over what enters the town sewers. Generally, the sewage will have the character of farmyard sewage except that it will be more dilute. Certain covered areas, such as the actual auction pens, can be strawed, thus giving a solid waste to be taken up and transported to a dump. External pens, which may be exposed to the weather, will most usually be hosed down after use and probably disinfected with chlorine, both as a public health precaution and to prevent nuisance. A special area may be allocated for the cleansing and disinfection of the cattle transporters, operations which should be carried out during the time the market is held so as to make clean trucks available for the transport of the animals after the market. Sidwick (1972) has estimated useful figures for the BOD loads likely to emanate from the various animals during their residence in a market. It should be noted that markets are probably the only occasions when sheep contribute to the waste disposal problem.

BIBLIOGRAPHY

Dept. Sci. Ind. Res., Water Pollution Research Laboratory, Notes on Water Pollution: No. 17 'Waste Waters from Farms', June 1962; No. 24 'Some Further Observations on Waste Waters from Farms', March 1964.

Papers published by the Institute of Water Pollution Control in its Journal, *Water Pollution Control:*

Little, F. J., 'Agriculture and the Prevention of River Pollution in the West of Scotland', 1966, Part 5, 452. (A useful general summary also containing information on silage liquors.)

Scheltinga, H. M. J., 'Aerobic Purification of Farm Waste', 1966, Part 6, 585. (Performance of oxidation ditches treating piggery wastes.)

Gibbons, J., 'Farm Waste Disposal in Relation to Cattle', 1968, Part 6, 623. (Explains in particular how methods of managing cattle have changed and the resultant effects on the waste disposal problem.)

Riley, C. T., 'Review of Poultry Waste Disposal Possibilities', 1968, Part 6, 627. (Contains many useful figures relating to quantity and composition.)

Pontin, R. A., and Baxter, S. H., 'Wastes from Pig Production Units', 1968, Part 6, 632. (A factual report on experimental oxidation ditches and their operation at a pig farm.)

Scheltinga, H. M. J., 'Farm Wastes', 1969, Part 4, 403. (Describes experimental oxidation ditches.)

Riley, C. T., 'Current Trends in Farm Waste Disposal', 1970, Part 2, 174. (A review of United Kingdom practice in manure disposal. Refers to copper as a contaminant of piggery waste. The discussion of this paper is of particular value.)

Wheatland, A. B., and Borne, B. J., 'Treatment, Use, and Disposal of Wastes from Modern Agriculture', 1970, Part 2, 195. (Contains figures supplementary to those quoted in Notes on Water Pollution Nos. 17 and 24. Describes experiments in treatment, including aeration and redox control. Mainly of research interest.)

McAllister, J. S. V., 'Collection and Disposal of Farm Wastes', 1970, Part 4, 425. (A comprehensive technical article on farmyard wastes quoting useful figures. Experience in N. Ireland, but refers also to European practice.)

Hepperd, R. Q., and Charlock, R. H., 'Performance of an Experimental High-rate Biological Filtration Tower When Treating Piggery Slurry', 1971, Part 6, 683.

Sidwick, J. M., 'Waste Treatment Plant at Pig Rearing Station, Cuyk', 1972, Part 2, 125. (Brief illustrated description of a full-scale oxidation ditch system; gives performance figures.)

Sidwick, J. M., 'Cattle Market Wastes', 1972, Part 5, 533. (An assessment of the problem with estimates of effluent strength.)

Institute of Water Pollution Control and University of Newcastle-upon-Tyne: 'Farm Wastes: A Symposium'. Proceedings published by the Inst. WPC, Maidstone, Kent, 1970. Contents:

Origins and nature of farm wastes, by K. B. C. Jones and C. T. Riley.

The problem of disposal of farm wastes, with particular reference to maintaining soil fertility, by C. Berryman.

The problem on the farm: animal health, by J. A. J. Venn.

Farm wastes: public health and nuisance problems off the farm, by T. H. C. Bartrop.

Water pollution prevention requirements in relation to farm-waste disposal, by H. Fish.

Sewers and sewage treatment, by J. R. Simpson and R. L. Hibberd.

Minimising the waste disposal problem in vegetable processing, by F. Barrett.

Minimising poultry waste problems, by C. T. Riley.

Minimising the problem with pigs, by C. G. Pointer.

Minimising the waste problem with cattle, by M. M. Cooper.

Building design, by J. B. Weller.
Building design and manure disposal, by J. C. Glerum, A. P. S. de Jong and H. R. Poelma.
Piggery cleaning using renovated wastes, by R. J. Smith, T. E. Hazen and J. R. Miner.
Land disposal and storage of farm wastes. 1. Planning and choice of system, by A. J. Quick.
Land disposal and storage of farm wastes. 2. Handling and distribution, by J. I. Payne.
Aerobic treatment of farm wastes, by K. Robinson, S. H. Baxter and J. R. Saxon.
Anaerobic treatment of farm wastes, by S. Baines.
Treatment of farm wastes, by H. M. J. Scheltinga and H. R. Poelma.

Other documents and papers referred to:

Jones, J. L., 'Slurry: Problem or Profit', *Country Life*, 1973, March 29, 873.
Ministry of Agriculture, Fisheries and Food, Advisory Leaflet STL 87, 'Farm Waste Disposal: Silage Effluent', March 1969.
Ministry of Agriculture, Fisheries and Food, Liscombe Experimental Husbandry Farm, Dulverton, Somerset, 'Grass Bulletin No. 2. Silage', Spring, 1972.
Ministry of Housing and Local Government, Welsh Office, 'Taken for Granted; Report of the Working Party on Sewage Disposal'. HMSO, London, 1970. (Paragraphs 335 to 361, inclusive, refer to agriculture and its waste products.)

Chapter 9

METAL FINISHING EFFLUENTS

S. W. BAIER

INTRODUCTION

The effluents from electroplating and related metal finishing processes (such as anodising of aluminium and phosphate treatment of steel prior to painting) normally contain appreciable quantities of various toxic materials—acids or alkalis, heavy metals such as nickel, chromium, copper, and zinc, and two specific types of salts which are considered to be particularly toxic, namely chromates and cyanides. Although there are probably still more toxic effluents from some specific industries those from plating shops can present quite a problem.

It is normally considered that these effluents should not be discharged into rivers, streams, or lakes if the total concentration of toxic materials should exceed more than a few parts per million— or more than 1 p.p.m., or even less, in the case of the most toxic constituents such as cyanides. Otherwise they will not only directly endanger the lives of the fish in the streams or the potability of the water if it should be used by a local authority as a source for drinking water supplies, but they may interfere with the whole biological balance of the watercourse. On the other hand where the effluents are discharged into a town's sewerage system, and consequently mixed with domestic sewage, the levels of toxic materials present may be about ten times greater, without causing much trouble.

In practice very much depends on the particular local conditions. When effluents are passed to the sewers of a large town where there is a high proportion of domestic sewage, the concentrations of toxic substances are so diluted that they cause little trouble at the sewage treatment works and the final concentrations in the effluents from

the works are low enough to be discharged into normal watercourses. On the other hand where there may be one or more reasonably large metal finishing shops in a small town, there would be much
less dilution by the ordinary domestic effluents, and the limits
allowable for toxic materials discharged into the public sewers
would have to be more stringent.

In the case of discharge to rivers or other watercourses, although
the safe limits must be very much lower, each case should really be
considered in the light of particular local conditions, the total
volume of the effluents flowing to the river, the size of flow of the
river and whether or not the effluent would be diluted to completely
safe limits within say a few feet of the point of discharge—also how
much contamination there may have already been higher up the
river and how much more might be added before it reaches the sea.
(*See also* p. 28.)

One obvious point which arises from these considerations is that
plating shops should never discharge their shop drains directly to
either rivers or sewers. They should at least feed them through
reasonably large 'balancing tanks' fitted with a suitable baffle
system, or even mechanical stirrers, so that the discharge is of
reasonably uniform composition and its toxic concentration does not
fluctuate from low levels to high levels as different processing operations are being carried out in the shops. Even more important, when
bulk solutions have to be discarded, they should be slowly bled at
suitable times of day into the balancing tank along with the discharge
of low concentration effluents such as rinse waters.

TYPES OF EFFLUENTS AND CONTAMINANTS

The processing procedure which occurs in plating and other metal
finishing shops is briefly as follows. The parts to be treated are first
immersed in a series of cleaning tanks containing alkaline solutions
to remove any grease or oil from the metal, then into acid solutions
to remove any scale or oxide films from their surfaces. The cleaned
parts are then passed through the particular treatment tanks, where
they may remain for a few minutes or up to an hour according to the
thickness of the coatings being produced. In the case of electroplating
there may be a series of coatings to be deposited over one another;
for example for chromium plating zinc die-castings (as used for
motor-car door handles and innumerable other items) they are first
plated with copper, then with nickel, and finally with chromium.

Frequently there are also post-plating treatments, *e.g.* chromate passivation after zinc plating. After each of these plating, anodising or phosphating treatments the parts have to be rinsed clean of the treatment solutions and thorough rinsing is also required after each cleaning step before the next operation.

Thus in any plating or metal finishing shop (apart from those dealing only with paint finishes) there is a series of effluents from the rinse tanks where the waters each become contaminated with particular chemicals, which may be classified into the following groups:

(1) Alkalis from alkaline cleaning solutions, *e.g.* caustic soda, sodium carbonate, phosphates, silicates, and sometimes also sodium cyanide.

(2) Acids from pickling and bright dipping operations, *e.g.* hydrochloric or sulphuric acid which are used for pickling steel, mixtures of nitric and sulphuric acid or nitric, sulphuric, and phosphoric acid which are used for brightening brass and aluminium respectively, while some acid pickles and dips also contain acid fluorides. The rinse waters from these operations, as well as being contaminated with these acids, also become contaminated with salts of the particular metals being treated, *e.g.* iron salts from steel work, copper and zinc salts from brass, or aluminium salts from aluminium parts.

(3) The specific chemicals from the plating, anodising, phosphating, or chromate passivation solutions:

(a) These solutions may be strongly acid like the sulphuric or chromic acid solutions used for anodising aluminium, the acid sulphate solutions which are sometimes used for copper and tin plating, or the very concentrated chromic acid solutions used for chromium plating. They may be moderately acid like phosphating or chromate passivating solutions or they may be almost neutral like nickel plating solutions or the bright chloride zinc or cadmium plating solutions which have recently been developed to replace the conventional cyanide solutions that are usually used to plate these metals.

(b) Other solutions may be strongly alkaline like stannate tin plating solutions and the caustic solutions used for etching aluminium, or much more frequently they are alkaline with high concentrations of cyanides as well, *e.g.* conventional zinc and cadmium plating solutions and cyanide copper, silver, and gold plating solutions.

As well as containing some acids or alkalis these solutions normally contain high concentrations of the various metal salts in the case of plating solutions, some aluminium in the case of the anodising or caustic etching solutions used for treating aluminium, metallic phosphates (iron, zinc, or manganese) in the case of phosphating solutions or dichromates in chromate passivating solutions.

The concentrations of these contaminants getting into the rinse waters will depend on the concentrations of the chemicals used in the processing solutions (which usually contain from 20% to 50% of dissolved chemicals) on the rate of flow of the rinse waters, and on the volume of work which is being processed (usually in proportion to the surface area of the articles being treated) and also to some extent on the shape of the articles treated, articles with re-entrant angles tending to 'cup' solutions and drag larger volumes over into the rinses.

As well as the more or less steady flow of these relatively lightly contaminated rinse waters, the shops have also to discard some of their processing solutions from time to time. Because of their high cost, the actual plating or anodising solutions are very seldom discarded, but are maintained in good working condition indefinitely by making small additions of fresh chemicals and by filtering and various other purification treatments. However, the various pickling, cleaning, or chromate post-plating solutions have limited lives and must be discarded at intervals of about 1 to 6 weeks, and these must be given special consideration.

A third source of toxic effluents is spillage and leakage from chemical pumps and filters, etc., which can get on to shop floors and find their way to the drains. They can sometimes cause more trouble than the actual rinse waters or even discarded solutions, for which most shops would organise systems of neutralisation. However, they can be dealt with by reasonably simple preventative measures. Rinse tanks should be immediately adjacent to all processing tanks so that work dripping with chemicals is not carried across floorways, and metal or plastic shields should be fitted across all gaps between processing and rinsing tanks, preferably sloping in such a way that the drips return to the processing solutions rather than to the rinses. Trays can be fitted under filters and pumps to collect leakage; which can often be returned from them to the processing solutions, or at least dumped with appropriate discarded solutions after special treatment. (In one case, in a small plating shop which had not been laid out well when originally planned, shallow trays filled with a neutralising solution (sodium hypochlorite) had been put alongside all plating tanks employing cyanide solutions so as to neutralise the cyanide drips; and this shop was thus able to reduce the cyanide level of their main effluent below that required for discharge to the local sewers.)

SYSTEMS OF DEALING WITH EFFLUENTS
The simplest system for dealing with the effluents would be merely to use sufficient rinse water to dilute the process effluents to a level at which they may be discharged without being harmful. Some thirty or forty years ago this was in fact the normal system in the majority of plating shops. Water was cheap, and plenty of rapidly flowing rinses ensured that the work was clean from pickling and alkaline cleaning before it entered the plating or other processing baths, and that the plated, anodised, phosphate, or chromate coatings were efficiently rinsed before drying and packing so that the metal finisher could be well pleased with his work. But, admittedly, at that time no one was particularly worried about the level of contamination unless it was doing some very obvious harm to streams, rivers, or local sewerage pipes or sewage treatment works.

However, as time has gone on, water has become much more expensive and is even a scarce commodity in some districts. As well as having to pay for the water supplied, some local authorities are also charging for water discharged to their sewers, and these charges may be scaled to increase with the level of contamination of the effluent and, in addition, there are also various maximum limits of concentration for specific chemicals. At the same time, the concentration of the plating and processing solutions has increased to secure more efficient or more rapid processing, and modern processes—especially with automatic processing plant—tend to become more and more complicated, with extra cleaning operations before plating or anodising and more and more actual plating stages in order to achieve better quality products. And lastly, of course, the whole level of industry has been increasing more and more rapidly and the metal finishing industry probably rather more rapidly than most.

The present position in England and Wales is that metal finishing shops which were discharging effluents to local sewers at certain rates in 1937, when the laws were modified, are still allowed to discharge at the same rates and at the contamination levels of that time. Thus shops which have not altered their output or complicated their processing treatments since then can continue on their old lines, at least for the present. However, if their output of work increases in size or complexity then they must find some way of only producing the same effluent discharge and contamination or perhaps arrange to reduce the contamination level proportionately to the increase in flow. However, if they should move their works, which many firms wish to do in order to expand, or have to do because local planning schemes force them to abandon sites in the centres of large towns, then the new shops must comply with the newer regulations applying to the particular area. These regulations are normally much more onerous than the original prescriptive rights, and they vary greatly from one district to another, being much more onerous in smaller inland towns (*i.e.* nearer the sources of river drainage areas) than in the larger towns, especially those reasonably close to the coast. For example, the level of contamination allowed in London is much higher than that allowed in Birmingham and it would be still more onerous in a small inland town—the actual levels allowed being partly dependent on the ratio of domestic sewage to industrial effluents and partly on the local river authority's rulings, since after purification at sewage works, the town sewage must finally discharge into a river basin.

CONVENTIONAL CHEMICAL TREATMENT

Except in very small finishing shops, which might be able to use sufficient water to render their effluents acceptable, it is normally necessary to install some method of treatment. The general principles are quite simple. First, the two most toxic contaminants, cyanides and chromates, if present, must be destroyed, cyanides normally by chlorination under alkaline conditions and chromates by reduction with sulphur dioxide or sulphites under acid conditions. After this, the effluents can be brought to an acceptable pH level, normally between 6 and 10, but sometimes between 7 and 9, at which levels they are acceptable for discharge to both sewers and rivers. At these pH values all the toxic or undesirable metallic radicals in solution, such as copper, zinc, cadmium, nickel, iron, and chromium (from reduction of the chromates) are precipitated as insoluble hydroxides and so can be removed by settling out or filtration.

For such treatment, it is essential to segregate the effluents containing cyanides, which are alkaline, from those containing chromic acid or chromates, which are either acid in the first place or have to be acidified before reduction.

The oxidation of cyanides by chlorine takes place in two stages:

$$NaCN + 2NaOH + Cl_2 \rightarrow$$
$$NaCNO + 2NaCl + H_2O \quad (1)$$

$$2NaCNO + 4NaOH + 3Cl_2 \rightarrow$$
$$N_2 + 2CO_2 + 6NaCl + 2H_2O \quad (2)$$

However, as the cyanate formed in the first stage is considered to be comparatively non-toxic* it is now seldom considered necessary to carry out the second reaction, especially as it can only be carried out satisfactorily by altering the pH after the completion of the first reaction and, in any case, it adds considerably to the cost of the chemicals required.

As shown, these reactions only take place when sufficient alkali is present, and so it is first necessary to add alkali—either caustic soda purchased for this purpose or a suitable alkaline effluent, e.g. from alkaline cleaners, if available. If chlorination is allowed to take place in neutral or acid solutions, the reaction can produce cyanogen chloride, which is a very poisonous gas. Where large volumes have to be treated the chlorination is most economically carried out with chlorine gas (from cylinders) but on a smaller scale it is much simpler to use sodium hypochlorite (bleach) solution which acts as a mixed solution of chlorine and caustic soda.

As well as destroying simple cyanides, like sodium cyanide, chlorination also decomposes any complex cyanides, such as copper, zinc, and cadmium, which are present in the effluents from these plating solutions, although it does not destroy complex nickel cyanides very easily, nor ferrocyanides which are still more stable. As a result of this decomposition of complex cyanides and the additions of alkali to raise the pH, all the toxic and undesirable metal radicals present are precipitated as hydroxides, as already explained.

---

* One estimate is that cyanates are 1000 times less toxic to fish than cyanides.

It is then necessary to reduce the pH after the chlorination treatment from its level of 10 or 11 down to the desired level of 8 or 9 before settling or filtering out the solids and finally discharging from the shops.

The treatment for chromates is to reduce them with sulphur dioxide, from gas cylinders for large scale treatment or with a sodium sulphite solution for simplicity for smaller scale treatment. The reduction only takes place in acid solutions and it is therefore usually necessary to add acid to maintain sufficient acidity throughout the reduction:

$$2CrO_3 + 3SO_2 \rightarrow Cr_2(SO_4)_3$$
$$2CrO_3 + 3Na_2SO_3 + 3H_2SO_4 \rightarrow$$
$$Cr_2(SO_4)_3 + 3Na_2SO_4 + 3H_2O$$
$$Na_2Cr_2O_7 + 3SO_2 + H_2SO_4 \rightarrow$$
$$Cr_2(SO_4)_3 + Na_2SO_4 + H_2O$$

After this reduction the effluents must be made alkaline (usually with caustic soda) to precipitate the reduced chromium (i.e. the chromium sulphate) as chromium hydroxide and bring the effluent to the desired pH range of 7 to 9, before finally settling (or filtering) and finally discharging.

Where both cyanides and chromates have been treated separately in this way, it is normal to mix the two effluents together before finally adjusting the pH to the required level and removing the solids. Effluent streams free from cyanide and chromate are normally also added at this stage or they may be fed into the appropriate stream before these chemical treatments, acid effluents, including those containing nickel salts (which are almost neutral) being added to the 'chromate' stream and all alkaline effluents to the 'cyanide' stream.

When it has not been possible to segregate effluents containing cyanides from those containing chromates there is no really satisfactory method of treatment, although ferrous sulphate treatment is sometimes employed, since ferrous sulphate does react with both chromates (reducing them to chromium sulphate) and with cyanides to produce insoluble Prussian blue (ferric ferrocyanide). However, the reactions are very slow and difficult to run to completion, and when the solutions are finally made alkaline, usually with lime, they give a very voluminous sludge which is difficult to settle out or filter effectively. Alternatively, it may be possible to use ion-exchange treatment, which will be discussed later; but it is rather expensive.

From this description of the reactions involved in these chemical methods for purifying effluents a number of points should become obvious:

(1) There is considerable expense for chemicals used in the treatments: chlorine or hypochlorite, sulphur dioxide or sulphites, caustic soda or lime and sulphuric acid.

(2) Quite close chemical supervision is necessary to maintain the required pH levels at each stage and to add the treatment chemicals in the right proportions and rates to neutralise the cyanide or chromate.

(3) The drainage channels in the plating shops have to be arranged so that the effluents (*i.e.* the rinse waters, floor spillage, etc.) from all solutions containing cyanides flow along separate streams from those containing chromates. Effluents containing other alkalis or acids may be fed into the 'cyanide' or 'chromate' streams respectively or may be kept separate too.

(4) Considerable plant and space are needed to carry out the treatments—'balancing' tanks to hold quite large volumes of effluents, suitable tanks for holding solutions which have to be discarded and bled into the rinse water effluents, storage tanks for solutions of the neutralising chemicals (hypochlorite, sulphite, caustic soda or lime slurries, sulphuric acid), treatment tanks with stirrers in which to carry out the chemical reactions and, finally, large settling tanks, to settle out the precipitated sludges or preferably, suitable filters, although so far no one seems to have developed really suitable filters to deal with large flows of effluents for this purpose without them being prohibitively expensive.

As regards chemical control, most plating shops of any reasonable size employ chemists to analyse and control their plating and processing solutions, and they can be given the extra task of controlling effluent treatments too, although frequently it would occupy a large part of their time. The other approach is to use as much automatic control as possible—automatic pH meters to control the flow of acids or alkalis to the treatment tanks to maintain the effluents at the correct pH levels at each stage and automatic redox potential controls for adding the chlorine or hypochlorite (to neutralise the cyanide) and the sulphur dioxide or sulphite (to neutralise the chromate).

For dealing with the volumes of plating effluents from large plating shops it may be necessary to carry out the chemical treatments on a continuous basis, the first pH adjustment being made as the effluent flows though a long baffled tank and the neutralisation of the cyanide or chromate in a second similar tank. The two 'neutralised' effluents (the cyanide and the chromate streams) then flow together into a third tank where the final pH adjustment is made (with either alkali or acid according to which of the two streams predominates). Finally, the effluent with the finely precipitated metallic hydroxides passes through a series of somewhat specialised settling tanks, sometimes including 'clarifiers' from which the settled sludge has to be removed from time to time and either shipped away as a rather liquid sludge (containing about 90% of water) to be dumped at some suitable site or, alternatively, passed through a plate filter to remove the bulk of the water to give a much dryer 'filter cake' which would be rather easier to dispose. Apart from the great expense of such plants they usually also require the employment of an operator and part time employment of a chemist as well as routine maintenance of the automatic pH and redox control equipment. Thus the cost of effluent treatment in this 'conventional' manner can be quite staggering. It is a cost for which the metal finisher gets no return in lower production costs or improved quality of his finishes. However, large works have had to accept the fact that their effluents must be treated to meet the requirements of the local authorities and have come to accept that it adds a considerable overhead cost onto their production costs.

For somewhat smaller shops it is possible to purchase 'packaged' chemical treatment units for treating plating effluents. Units to handle 2000, 4000 or 10 000 gal of effluent per hour are advertised. Presumably unless larger quantities of effluent than these are to be dealt with, their capital cost would be less than having to design, purchase, and install individual treatment plants to suit the requirements of the particular shops. The units apparently contain the treatment tanks and stirrers, the pH and redox control instruments and the dosage equipment which they control, but presumably pumps to feed in the effluents and suitable sludge settling tanks may have to be added. They still entail some labour to make up stock solutions of neutralising chemicals, a certain amount of routine maintenance of the instruments, and occasional chemical analysis to confirm that they are working satisfactorily.

SIMPLIFYING AND REDUCING COST OF TREATMENT

For small shops and even for large ones the important question that arises is how far can the amount of chemical treatment be reduced? To this end, the following specific questions need to be answered:

(1) Can the volume of effluent be kept down so that it will only be necessary to have a plant to deal with smaller volumes (and at the same time reduce the volume of the final 'purified' discharge, especially where local authorities make charges on a volume basis for discharges) even though the actual amount of contaminants passing into the rinse waters (and the amount of chemicals needed to neutralise them) is not reduced? In dealing with batch treatments in smaller shops, could the volume of effluent collected for treatment each day be reduced from 5 or 10 000 gal to 1 or 2000 gal; or could it perhaps be reduced to a single batch of 5 or 10 000 gal to be treated at weekly intervals?

(2) Can the total amount of contamination getting into the rinse waters be reduced and thus save a proportion of the cost of the chemicals required for the treatments, at the same time probably reducing the volumes of effluent to be dealt with?

(3) Can any materials be collected from the treatment processes which could either be returned to the processing tanks in place of some or all of the normal 'make-up' chemicals or alternatively in the form of saleable waste products containing valuable metals or chemicals that could be used by other industries. Also is it possible to recover some of the water from the rinses in a pure form or in a pure enough form to be re-used in the shops? If any of these can be done, it would at least offset some of the chemical or labour costs entailed in having to treat the effluents.

(4) Lastly, is it possible to avoid using the really toxic chemicals, such as cyanides and chromates which need the more complicated and expensive chemical treatment for their neutralisation?

The general answer is that all these savings can be achieved to a greater or less extent by careful planning, sometimes merely by installing extra tanks, at other times by installing special equipment, sometimes by altering the plating or processing procedure and often by altering the rinsing techniques.

ION-EXCHANGE

For the first type of saving, *i.e.* carrying out chemical treatments on small volumes of concentrated solutions instead of on large volumes of dilute solutions, ion exchange is the obvious method. In fact, effluent treatment by ion exchange is merely a method of concentrating and partially separating the contaminating radicals so that they can be precipitated out more easily by classical chemical methods. However, it is usually a relatively expensive method. Its big advantage is that it also recovers most of the rinse waters in a highly purified state for re-use in the shops. Therefore, where water is expensive or scarce and especially where charges are also made on the volume of the final discharge, it may provide the cheapest method of effluent treatment, and in a few very large plating shops where these conditions obtain it is probably the most practical method. In smaller shops it has also been used because of its convenience as a method when it has not been possible to segregate rinse waters into two separate 'cyanide' and 'chromate' containing streams.*

The principle is that the effluents are passed first through a bed of cation exchange resin where contaminating metal ions (copper, nickel, zinc, cadmium, etc.) are exchanged for hydrogen ions from the resin. The discharge from this exchange unit, in which all the anions (chromate, cyanide, chloride, sulphate, etc.) are present as acids, is then passed through a bed of anion exchange resin where they are exchanged for —OH ions from the resin, the final discharge being pure (de-ionised) water which is passed back for re-use in the plating shop. When the ion exchange resin beds are saturated with the metal cations and acid anions respectively, the flow has to be stopped and the resins regenerated. The point of saturation is indicated by a conductivity meter which indicates a sharp rise in the conductivity of the deionised water discharge when the resins start to get saturated.

Sometimes two separate different types of anion exchange beds are used, the first to deal with strong acid anions and the second with weak acid anions. At other times 'mixed-bed' exchangers are used instead of separate cation and anion exchange beds.

The ion exchange resins are regenerated to their original 'hydrogen' or 'hydroxide' forms by passing a relatively strong solution of acid (usually hydrochloric) or caustic soda through the respective cation

* Details of a large installation using ion-exchange were described by Pearson and Parker, *Trans. Inst. Metal Finishing* (1961), **38**, 159, and is discussed in R. Weiner's book *Effluent Treatment in the Metal Finishing Industry*, Robert Draper, Teddington, 1963, together with a description of a small ion-exchange installation used in a fairly small general plating shop.

or anion beds and at the same time the contaminating cations (copper, nickel, zinc, cadmium, iron, etc.) and anions (cyanide and chromate) that have been collected on the resins are recovered as concentrated solutions in the relatively small volumes of acid or alkaline eluates. These eluates can then be treated by the conventional chemical 'neutralisation' methods much more effectively and with much smaller plant than is possible when the treatments have to be carried out directly on large volumes of quite dilute effluents. Also the precipitated metal hydroxide sludges can be collected on ordinary filter presses and the clear filtrates, which only contain harmless sodium salts such as chlorides or sulphates, can be discharged direct to sewers or perhaps after some dilution even to river courses.

Apparently, as mentioned above, it is quite practicable to 'neutralise' a mixed alkaline eluate containing both cyanides and chromates when dealing with small volumes of these concentrated eluents, either by chlorination of the cyanides followed by acidification and reduction of the chromates and final neutralisation to pH 8–9 to precipitate the chromium as hydroxide, or by treatment of the solution with ferrous sulphate, which reduces chromates and precipitates both the cyanide as Prussian Blue and the reduced chromium as hydroxide; and so ion exchange would seem to be one of the best methods of dealing with such mixed effluents when shops have not been planned to segregate their cyanide and chromate effluents into separate streams.

For large-scale shops it is normal to have two pairs of ion exchange units so that one pair can be in use while the second is being regenerated. For smaller scale shops a single pair may suffice and may be regenerated once or twice a week at idle periods, or holding tanks might be practicable to hold the rinse water effluents during regeneration periods. The two types of eluates would be collected in two tanks, to which appropriate discarded solutions could also be added, and they would be held till their volumes were large enough for the chemical treatments and filtration to be carried out on reasonable sized batches.

There are, however, a few snags which must be guarded against when using ion-exchange treatment in plating shops, which were not realised when many of the earlier plants were installed. Firstly, it is essential to pass the effluents through filters before entering the ion exchange beds, otherwise the ion exchange resins can become clogged rapidly with fine solids and this will greatly reduce their efficiency.

Secondly, organic materials in the solutions, such as the organic brighteners used in nickel, copper, and zinc bright plating solutions, must be removed before ion exchange treatment, otherwise they are absorbed on the surface of the resins and 'poison' them. Such materials can be removed quite simply by incorporating activated carbon on the filters prior to the ion exchange treatment. If these precautions are not taken the exchange resins quickly become so inefficient that they have to be scrapped and replaced, which is very expensive and should normally be necessary only after a period of several years. Lastly, it has sometimes been found that the exchange resins require regenerating after very short periods (a matter of hours rather than days) because of the hardness of the water supply, the amounts of calcium and magnesium being removed by the ion-exchange treatment exceeding that of the actual contaminating metals from the plating shop rinses. Such a situation can be put right by installing an extra ion exchange unit (a simple water-softener type) through which all the 'raw' water passes before being used for the plating shop rinses.

In general, the total volume of water being passed through an ion exchange unit is of little consequence. In many ways the more dilute the rinse waters, the more efficient the process—and there is no actual extra consumption of water. It is merely a matter of circulating it through the exchange units and back through the rinses more quickly.

However, it should be realised that ion exchange does not reduce the amount of chemicals needed for treating the effluents and, in fact, it increases the chemical costs because further chemicals are required for regenerating the ion exchange resins. Also the exchange plant is relatively expensive. Its value compared with direct chemical treatment is purely one of reducing the amount of incoming water, the volume of the final discharge and of making the chemical treatment rather simpler and more efficient and so the expense of ion exchange treatment is only worthwhile if it can be offset by economies in one of these directions.

It is claimed that ion exchange can be used to recover chemicals for re-use in the plating shops or as saleable residues. Theoretically this is so. For instance, if the rinses from nickel and chromium plating are kept in separate streams and treated individually in ion exchange units, then the eluate from the first would be an acid solution of nickel chloride or sulphate which can be returned directly to the nickel plating tanks, but only at times when they require additions of acid as well as nickel; otherwise the recovered solution has first to be neutralised with nickel carbonate. However, in the second case, where the exchange would be by anion instead of cation exchange, the eluate would be a solution of sodium chromate, which could not be used in chromium plating solutions, unless it was first put through a cation exchanger to convert it to chromic acid. Recovery from rinses from cyanide plating solutions would be still more difficult to achieve. Part of the metals might be collected as acid solutions of chlorides or sulphates from a cation exchange unit and part as an alkaline eluate containing cyanide complexes of the metals, which might be returnable to the plating solutions. However, it is quite economic and practicable to purify two particular types of solution by ion exchange—namely, chromic and phosphoric acid solutions— to avoid having to completely, or partially, dump these concentrated solutions at intervals.

Engineering Chromium Plating Solutions

In solutions used for obtaining thick chromium deposits ('heavy' or 'hard' chromium) for engineering purposes, there is a gradual build-up of iron and trivalent chromium and this eventually makes them unusable. At this stage they are normally completely or partially discarded and made up afresh. However, it is quite practicable to pass such solutions through a cation exchange unit using a somewhat special type of resin which will withstand the oxidising action of chromic acid solutions containing between 100 and 150 g/litre of CrO_3. This extracts the iron and trivalent chromium and any other heavy metal contaminants to give a pure chromic acid solution, and the treatment actually regenerates the chromic acid which is combined with the iron and trivalent chromium, as well as 'collecting' the free chromic acid present. For successful treatment it is, how-ever, necessary first to dilute the plating solutions with about an equal volume of water to reduce their chromic acid concentration, but there is usually sufficient evaporation in the plating tanks to allow the greater volume of the more dilute solution to be returned direct to the plating tanks.

Decorative Chromium Plating Solutions

Normally these do not get so seriously contaminated with iron, trivalent chromium, or other heavy metals, but obviously they could be treated in a similar way if they did.

Chromic Acid Anodising Solutions

These are rather more dilute solutions of chromic acid in which large amounts of aluminium and trivalent chromium build up as they are used. They also become inoperable when the concentrations of these metals become too high and they can be regenerated by ion exchange in the same manner as chromic acid plating solutions, but without any need to dilute before the treatment.

Phosphoric Acid Solutions (for Pickling Iron and Steel)

These would normally have to be discarded when they become saturated with iron and their acid content drops correspondingly; but like chromic acid solutions they can be purified by ion exchange to remove the iron and recover the whole of the original acid content. With expensive acids like chromic and phosphoric acids, ion-exchange becomes well worthwhile, even though fairly large amounts of hydrochloric or sulphuric acid are consumed in regenerating the ion exchange resins. As well as recovering the acids for re-use, the other important aspect is that these acids are kept out of the shop effluents altogether and only the acid eluents from regenerating the ion exchange resins have to be treated as an effluent.

COUNTER-FLOW RINSES

A much more logical approach to the effluent problem would be to try to prevent a large amount of the contamination getting into the rinse waters which have to be discharged from plating or metal finishing shops. At the same time the volume of the effluents should be reduced and a substantial proportion of the processing solutions saved. This can be achieved to a greater or less extent by the use of counterflow rinses of one sort or another and it is largely a matter of adding a number of extra rinse tanks, altering the water piping for feeding them and the general techniques of rinsing.

This is the logical first approach and only after everything possible has been done in this direction of arranging counterflow rinses should one plan how much chemical treatment will really be necessary to ensure the required level of purification of the outgoing effluents. In fact, it is probably true that the use of ion-exchange as a method of treatment for plating shop effluents has only been necessary in cases where it was impossible to install sufficient counterflow rinses, and something had to be done without drastic alterations to the plating plants to meet urgent requirements to prevent pollution. These cases arose both with large-scale automatic plating plants, where it would have become a major engineering operation to increase the number of rinsing operations after each processing tank, and in smaller scale general plating shops where it would have involved an almost completely new lay-out of the shop to find the space for the necessary alterations to their rinsing procedures.

Theoretically, by using sufficient counterflow rinses after a plating operation it should be possible to use only as much water for the rinses as can be returned to the plating tanks to compensate for evaporation losses. If the evaporation loss was infinitely small it

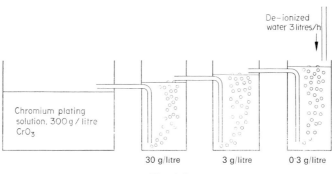

FIG. 9.1

would obviously require an infinite number of counterflow rinses, but most plating solutions are used warm and require to be topped up with quite appreciable volumes of water or solution each day to make up evaporation losses. Theoretically, with such a system, all the rinse water could be counterflowed back to the plating tank and there would be no effluent at all to dispose of (*see* Fig. 9.1).

As an example, take a 200 gallon tank of chromic acid plating solution containing 300 g/litre of chromic acid which is plating a load of work with a surface area of 5 ft^2 every 5 min. Assume that each square foot of surface was wet with 5 ml of plating solution when the work was passed from the plating tank to the first rinse. Then each load would transfer 25 ml of the plating solution, containing $7\frac{1}{2}$ g of chromic acid—*i.e.* 90 g/h to the first rinse tank. If there were three rinse tanks in series with the rinsing water fed to the last and counterflowed back through the first two and 3 litres of water per hour was fed into tank No. 3, then one could allow this

3-litre flow to run directly into the plating tank, as it would probably be about sufficient to make up for evaporation. Calculations show that the concentration of chromic acid in the first rinse tank would climb up until it reached a steady concentration of 30 g/litre, the concentration in the second tank to 3 g/litre and in the third to only 0·3 g/litre. If this final concentration is considered low enough to be allowed to dry on the work without spoiling the appearance of the finish, as it might easily be, then the whole effluent problem from the chromium-plating operation would be eliminated. It should be noted that the concentration of the contaminants in the final rinse water clinging to the work before drying it off can be very different from that which would be allowed in an effluent discharge— 0·3 g/litre chromic acid—*i.e.* a 300 p.p.m. solution would be nearly 100 times too great for discharge but might be quite acceptable as a thin film to be dried on the work. One should also realise that with conventional single- or even double-stage flowing rinses the final rinse water is by no means uncontaminated and might easily contain up to 1 g/litre of contaminants. A similar calculation would apply to rinsing after nickel plating where the concentration of nickel salts would be of about the same order.

The calculations used for this example assume that the bulk of the plating solution has been allowed to drain off the parts before they are transferred from the plating tank to the rinse, and from one rinse to the next and that the plated parts are allowed to remain in the rinses long enough for the solutions clinging to them to become mixed completely with the rinse waters. If the 'drag-out' of solution on the work was greater than in the example, if the 'efficiency' of mixing in the rinses was less than 100%, or if the initial plating solutions were more concentrated, then the flow of the rinse water would have to be increased proportionately and it might be difficult for the flow back to the plating tank to be accommodated by the rate of evaporation from the tank.

However, it is obvious that the practical success of a counterflow rinse system depends on making sure that sufficient draining time is allowed and that efficient mixing occurs in the rinses. The first is not difficult with an automatic plating plant where it can be arranged for a pause for draining back into the plating tank when the work is lifted out of the solutions and before being passed forward into the rinses. With manual working, operators would have to be trained to make comparable pauses for drainage when they lift work from one tank to another and it is often helpful to fit 'holding bars' above the plating tanks on which each jig of work can be hung to drain while the next is lifted out and then to pass the first on to the rinse and so on. Similarly, it is important to obtain maximum 'mixing' in all the rinse tanks, which can be achieved by circulating the rinse waters through effective paths in the tanks and/or by providing air agitation in the rinse tanks, and allowing sufficient time in the rinses.

Even when no counterflow rinsing is employed it becomes obvious that the same simple techniques of proper drainage and proper mixing in the rinse tanks could still considerably ease the problems of effluent treatment. More efficient drainage, especially on manual plants, might easily reduce the amount of contaminant in the rinse waters by 50%, and efficient mixing in the rinse tanks might easily reduce the volume of rinse water necessary to rinse the work satisfactorily for drying-out by about the same percentage. Thus these measures might easily halve the amount of chemicals needed for chemical neutralisation treatment and halve the volume to be treated, making the treatment plant required that much smaller.

However, the employment of a complete counter-flow rinsing system, which could theoretically completely eliminate the discharge of any effluent as has been described, would involve considerable practical difficulties. Firstly, at least three or perhaps even four or five rinse tanks might be required after each plating tank instead of the usual one or two. In a manual plating shop this would increase the shop area required by about 20-30% for each plating process, or even more in some small shops were common rinse tanks are sometimes employed to rinse work from two or more plating tanks. Also it would about double the present labour time for unloading and rinsing the plated work, as the operator would have to lift the work in and out of a series of rinses instead of in and out of one (or perhaps two) as at present. Also there would be the necessary delay in holding work over the plating tanks and rinse tanks to allow it to drain and perhaps in moving the jigs up and down in rinse tanks to get more efficient mixing instead of leaving them static. In the case of automatic plants extra tanks would have to be added after each plating stage; this would probably only increase the length of the plants by 10% or 20% but it would involve the installation of one or more 'transfer' operations after each plating operation. Also the water supply to any counterflow rinse system must be pure deionised or even distilled water, as discussed later. The economics would have to be calculated for each particular installation and compared with the savings gained by eliminating or reducing the cost of effluent treatment.

A second objection to counterflow systems is that impurities which accumulate in the plating solutions are perpetuated, whereas, when the rinse waters run to waste, these impurities get 'bled off', as it were, along with the plating solution dragged-out, so that the impurity concentrations eventually level out at some acceptable value. However, this seems to be debatable because it implies that it is economic to allow a fair proportion of the actual plating chemicals to run to waste in order to take away a smaller percentage of unwanted impurities with them, rather than to use some sort of purification system for the plating solutions themselves to eliminate the impurities. This emphasises one obvious necessity: very pure water, such as deionised water, must be used for counterflow rinse systems, otherwise the chemical radicals present in the water would accumulate and remain for ever in the plating solutions, *e.g.* calcium from the water would lead to a rapid build-up of sparingly soluble calcium sulphate in nickel plating solutions which would lead to serious roughness of the nickel coatings; or the chloride from the water would accumulate in chromic acid plating or anodising solutions, which become inoperable with even quite small chloride contents. In the case of nickel-plating solutions, especially those for bright nickel plating, the drag-out into rinse waters is in any case not sufficient to eliminate the accumulation of metallic or organic impurities (such as break-down products from organic brightening agents) and auxiliary purification treatments already have to be installed. These could easily be increased in capacity to take care of the extra impurities returned to the solutions with the concentrated rinses. Similar considerations would probably apply to zinc and many other plating solutions, but in the case of decorative chromium plating solutions, where there is no known direct chemical or electrochemical treatment to remove heavy metal contamination, it might be necessary to purify occasionally by ion-exchange treatment as recommended for 'hard' chromium-plating solutions.

EMPLOYMENT OF DRAG-OUT TANKS
In practice it is not necessary to go to the full extent of a 'complete' counterflow rinse system in order to obtain considerable reduction

in the contamination of the outgoing rinse waters and appreciable return of the plating chemicals to the plating tanks and thus to achieve a useful easing of the effluent treatment problem. The simplest system is to use a 'drag-out tank' before the first rinse. This is a 'static' rinse tank, which is filled with water in the first place. The work is rinsed in this and then passed through the normal single (or double) rinse tanks. This 'static' rinse should also be air-agitated to give efficient rinsing—*i.e.* mixing of the concentrated solution layer on the work with the static solution in the rinse and should preferably be made up with de-ionised water to prevent calcium, magnesium, chlorides or sulphates, etc. from accumulating in the solutions. Drainage periods before the work is transferred from the plating solution into the static rinse and on withdrawal from the static rinse to the following flowing rinse should also be used. As the plating solutions require topping up for evaporation losses, solution from the static rinse is used instead of water and the static rinse topped up with deionised water. Where this system is used it would seem important to top up the plating solutions at fairly frequent intervals from the drag-out tanks, otherwise the concentration of solution in the drag-out rinse would become rather concentrated and excessive plating chemicals would still remain on the work being carried to the flowing rinse.

Where the rates of evaporation are not great enough to put sufficient amounts of a static rinse back into the plating solutions, they may be concentrated first and it has been claimed that it is even practical and economic to concentrate drag-out solutions in vacuum or low-pressure evaporators, the condensate of pure water being returned for re-use in the rinses. It is claimed that this can be practicable, particularly for rinses from chromic acid and cyanide plating solutions which cannot be concentrated by simple evaporation without decomposition. Although it seems likely that the plant would be very expensive, elimination of the necessity for conventional chemical treatment to neutralise the two most awkward contaminants (chromic acid and cyanides) is obviously very desirable.

INTEGRATED WASTE TREATMENT

Another recent approach is the 'Integrated Waste Treatment' system or Lancy process. In this system, instead of flowing all the rinse waters into a common stream (or into two or three separate streams to isolate cyanides from chromates) a chemical neutralisation stage is introduced immediately after each plating bath, the work being 'rinsed' in the neutralising solution before being passed to the normal rinse tanks. Although full details of the neutralisation solutions have not been disclosed it can be assumed that they are fairly similar to those used for conventional chemical neutralisation —*i.e.* fairly simple alkali solutions to neutralise and precipitate nickel or copper as hydroxides, or perhaps as carbonates, from acid or neutral solutions, hypochlorite solutions to decompose cyanides, and sulphite and alkali to reduce chromates and precipitate chromium hydroxide. Relatively weak solutions of these neutralising chemicals are used in the treatment tanks following the plating tanks, sufficiently concentrated to neutralise all the chemicals dragged out on the surface of the plated work but too weak to have any harmful corrosive effects or to cause tarnishing or discoloration of the plated coatings. When the plated work passes from the (integrated) treatment tanks to the final rinses any of these weak solutions, with their suspended hydroxide precipitates, left on the surface is washed off and the work left thoroughly well rinsed and ready to dry. The only contamination of the final rinses would be traces of the neutralising chemicals, which are fairly harmless and would be acceptable in the final discharge together with some suspended precipitates which can be allowed to settle out in settling tanks before the rinses are discharged.

The 'neutralising' solutions in the treatment 'rinse' tanks are kept in continuous circulation to and from large 'storage' tanks containing reasonably large volumes of these specific solutions to which the neutralising chemicals are slowly fed to replenish the quantities used up in 'rinsing' the work. These tanks are designed so as to act also as settling tanks for the precipitated hydroxides formed in the 'rinse' tanks. Where two or more plating tanks carrying out identical processes are employed in the shops the 'integrated' treatment tanks are all circulated through a single 'storage' tank, *i.e.* there need only be one 'storage' tank for all the nickel-plating neutralising rinses, one for all the chromium-plating 'rinses' and so on. Presumably the concentration of the chemicals in the storage tanks must either be controlled automatically by pH and/or redox controllers or by direct chemical analyses. The hydroxide sludges collected in the storage tanks are removed from time to time, *e.g.* at monthly intervals, and collected on filter presses or other filters. If the sludges are of little or no value they can be discarded, but the latest information indicates that where they contain valuable metals these may be reclaimed. In particular the precipitated nickel hydroxides are free from any serious contamination and can be dissolved in dilute acid for adding to plating tanks as replacement chemicals. It is not clear whether or not any flocculating agents are added to the treatment solutions to facilitate settlement and filtration of the 'hydroxide' sludges. It is claimed that much of the rinse water used in the system is pure enough to be recirculated to various rinse tanks in the plating shops and the system can therefore also give very large savings in water consumption.

MODIFICATION OF PLATING SOLUTIONS

First, to consider how the plating solutions themselves might be modified to ease the problem, it is obvious that the weaker the solutions are, the smaller the amount of contamination reaching the effluents. However, except in the case of chromium plating, there is not much which can be done in this direction without making the plating slower. For decorative chromium plating, solutions containing 500 g/litre chromic acid (*i.e.* 5 lb/gal) are sometimes used, particularly in smaller plating shops, because they are slightly less critical to contamination by iron and other heavy metals and to small changes in sulphate content than is the standard solution containing only 250 g/litre chromic acid. They can be used at rather lower current densities and voltages—*i.e.* one can plate large articles with such solutions using rather smaller rectifiers or generators—but obviously they will double the chromate contamination of the effluent. From this point of view it would obviously be better to use the standard 250 g/litre solution or better still to use one of the proprietary weaker chromic acid solutions which work perfectly well with only 150 to 200 g/litre of chromic acid.

To what extent are cyanide plating solutions necessary? For zinc plating, 'low cyanide' solutions containing only about 1 oz/gal of cyanide or less are now available and are nearly as easy to use as conventional high cyanide solutions. With care the use of such solutions might even make it possible merely to neutralise the effluents with acid to pH 8 or 9 to precipitate the zinc (perhaps also adding a little ferrous sulphate) to achieve a concentration of about 10 mg/litre of cyanide in the discharge. There are also pyrophosphate and cyanide-free alkaline zincate solutions available, which could eliminate cyanide altogether, but they are perhaps rather difficult to control and do not seem to have become popular. However, very recently bright, slightly acid or almost neutral zinc chloride solutions have been developed and similar cadmium plating solutions, which should make it possible to eliminate cyanides altogether for plating these two metals.

In the case of copper plating from cyanide solutions, there is so far no alternative to using copper cyanide for the first stage when nickel/chromium plating zinc die-castings. It has been suggested that work was being carried out to develop a pyrophosphate copper strike solution to replace the cyanide solution but so far this has not materialised. However, it should be possible to use a relatively weak cyanide strike solution and then to build up further thickness of copper from a bright acid copper or preferably from a bright pyrophosphate copper solution, before nickel and chromium plating. For copper plating steel it is sometimes quite practical to use an initial starting coat of nickel instead of 'cyanide copper' and then to plate the main copper layer over this from an acid or pyrophosphate copper solution.*

Recent research promises considerable progress in developing chromium plating solutions which are based on chromium chloride instead of chromic acid. A few years ago the BNFMRA patented solutions containing chromic chloride and dimethyl formamide for decorative chromium plating, and following further improvements it is now hoped that proprietary chromium plating solutions based on this work will soon be available to the plating industry. They should give considerably less trouble with effluent disposal than the present chromic acid types of chromium plating solutions.

The use of cyanides in alkaline cleaning solutions should be avoided. Admittedly, cold cleaners containing high concentrations of cyanide are very convenient to use, but there are equally effective hot alkaline cleaners which contain no cyanide, and proprietary cyanide-free cold cleaners are also available.

---

* However, with nickel/micro-discontinuous chromium-plating systems, which have recently been introduced to give better outdoor corrosion resistance than conventional nickel/chromium plating, it has been stated that the corrosion resistance is considerably poorer if an initial nickel 'strike' is used on the steel instead of a satisfactory 'cyanide' copper strike.

CONCLUSIONS

The logical way of dealing with the problem of reducing the contamination in effluents discharged from plating and other metal finishing shops, especially in the case of smaller plating shops, is first to examine the possibility of using processing solutions which are free from cyanides or chromates or alternatively to see if the concentration of the cyanides or chromic acid in the solutions could be reduced to at least half the present concentrations. Next, the amount of toxic chemicals entering the rinse waters should be reduced by the use of drag-out tanks or a suitable counterflow rinse system—either of which would often save appreciable amounts of chemicals being lost from the processing solutions. This would reduce the costs of replenishment chemicals needed for the plating solutions and there would also be appreciable savings of water.

If these measures would reduce the contamination of the rinse waters by cyanides and/or chromates sufficiently, well and good, since any other rinse waters from alkaline cleaning, acid pickling solutions or other less toxic plating solutions could then be dealt with simply by adjusting the pH with acid or alkali, to the acceptable discharge level of pH 8–9, the metal content of the solutions being precipitated at the same time and removed from the effluent by settling or filtering before discharge. If the chromic acid or cyanides could not be removed sufficiently then it would be a matter of examining which streams of rinse waters should be sent through treatment tanks where hypochlorite and alkali would be added in the case of cyanide containing rinses, or sulphite in the case of chromate containing ones. Only if the size of the effluent flow would be too great to deal with in a simple form like this would it be necessary to consider the use of a full chemical treatment plant, ion-exchange, etc. or perhaps the integrated waste treatment system after certain plating tanks.

Finally, the treatment and neutralisation of any processing solutions which have necessarily to be discarded must be considered. Normally, plating solutions never have to be totally discarded, although sometimes if they get in very bad condition they may have to be partially discarded before fresh chemicals are added to bring their composition back to the correct formulation. However, both pickling solutions and alkaline cleaning solutions and a few special solutions, like chromate passivation solutions, do have to be discarded and made up afresh at intervals varying from a few days to several weeks. The obvious course in the absence of a full chemical treatment or ion exchange plant would be to treat each solution as a separate batch (collecting in separate storage tanks if necessary to get a sufficiently large batch to treat), adding the requisite weight of chemicals to neutralise them. The batches would then be passed through settling tanks or filters to remove the solid sludges and finally discharged after checking that the pH was satisfactory. Normally it would be best to calculate the amount of chemicals required on the basis of a chemical analysis of each solution before treatment. Very convenient test sets are now available for simple colorimetric analysis of small concentrations of metals, cyanides and chromates in effluent solutions or for residual amounts of hypochlorites or sulphite left behind in effluents after treatment.

Chapter 10

THE TANNERY EFFLUENT PROBLEM
S. WOLSTENHOLME

Leather manufacture requires immense quantities of water. One square foot of light leather may well have passed through the equivalent of 10 gallons, in other words 400 times its weight of water. Since practically all the water used ends up as effluent, it will be seen that a tannery producing 10 million cubic feet of light leather per year is turning out a volume of effluent equal to a town of about 10 000 population. When we also take into account the polluting strength of the effluent then the population equivalent is nearer to 75 000.

Some leathers, for example heavy sole leather which is vegetable tanned, require less water because the object is to fill the hide as much as possible with tannins and other substances to give weight, toughness and solidity. Light leathers for shoe uppers and clothing, which are usually chrome tanned, need large quantities of wash waters to reduce as far as possible the amounts of residual salts which would cause trouble in shoe and garment processing and in wear. In fact, the better the quality of the leather the greater the quantity of water that will have been used in its manufacture.

COMPOSITION OF EFFLUENT
Hides and skins pass through many liquors, each quite different in chemical composition and each playing its part in the conversion of an unstable fibrous natural protein into a relatively stable non-putrescible leather. It is not possible to go into the details of leather technology here but we will consider these various liquors and list the various chemical constituents.

BEAMHOUSE LIQUORS

Soaks: These start off as relatively clean water but after use contain a fair amount of salt and soluble protein matter. In addition, dirt and pieces of skin flesh are floating about.

Limes: When made up these may contain up to 8 % lime, up to 2 % sodium sulphide and up to 2 % salt. After use these chemicals may be diluted and a little of the sulphide lost but in addition considerable quantities of protein matter and a small amount of ammonia will be in solution. The lime liquors are the worst polluting liquors of the tannery.

Bates: These are buffered enzyme solutions (pancreatin, etc.) and when discharged contain enzyme, calcium, and ammonium salts with soluble protein.

Pickles: These contain dilute sulphuric acid and 5–8 % of salt.

TAN LIQUORS

From Sole Leather Tanneries

Waste Vegetable Tan Liquors: These contain considerable quantities of natural polyhydroxy acids which absorb oxygen from solution readily and in so doing become darker in colour.

These liquors are in effect the chemical-reacting liquors responsible for the conversion of hide into leather, but they may only account for 15–30 % of the total effluent.

From Light Leather Tanneries

Waste chrome liquors will contain up to one per cent of chromium in the form of basic chromium sulphate together with neutral salts such as sodium sulphate and sodium chloride.

OTHER LIQUORS

Waste Dye Liquors (light leather): These contain unabsorbed dye-stuff, emulsified fatty matter and, sometimes, surfactants.

In addition, the hides (or skins), after having been removed from one reacting liquor, are given copious washes with clean water to remove residual chemicals before the skins enter a different liquor. These wash liquors amount to 70–85 % of the total effluent. As far as polluting properties go they range from very dirty lime wash liquors to relatively clean water which has been used for washing chrome leather after the leather has been neutralised.

So the final effluent is a constantly varying liquor made up from different quantities of the constituent liquors which may be discharged at different times during the working day. If the tannery makes several types of leather (which is not uncommon), then the pattern of discharge may vary from day to day. Generally, soak and lime liquors tend to be disposed of early in the morning since the skins require mechanical treatment before entering the pickle or tan liquors later in the day.

SCREENING
Tannery effluent contains large amounts of suspended matter. Much of this consists of lumps of 'stringy' flesh and hair. This, together with lime and fatty matter, will build up a thick layer on almost any surface; it also plays havoc with pumps. Bar screens need almost full time attention even when placed 2 in apart. This is due to the pieces of flesh wrapping round and thus blocking up space between the bars, sometimes in minutes. Satisfactory screening may be done with a Longwood flock-catcher. This is a half-cylindrical, stationary, perforated sheet, which is swept continually by revolving brushes. The original design, which was intended for the collection of wool from woollen trade effluents, was not satisfactory for tannery use, but after modifications it has proved very suitable for the very difficult job of dealing with tannery effluent. The following alterations were made to the flock-catcher:

(1) The drive needed to be more robust, so the epicyclic gears were replaced by a worm gear box.

(2) The original vegetable bristle brushes (like a yard brush) filled up with fleshings, and the alkalinity caused the bristles to break off. In order to combat this the four arms, each with its brush, were replaced by five arms, four were fitted with rubber squeegees (3 in × $\frac{3}{4}$ in) and one with double, nylon-bristle strip brushes.

(3) The modified scraper rubbers and brushes were made so that they did not extend to the full length of the flock-catcher and were kept about 4 in short of this in order to avoid jamming by hard lumps of foreign matter, such as wood blocks, etc., which occasionally float down.

(4) It was soon found that fine pulped hair, fat, lime, and fleshings built up a layer on the underside of the perforated sheet, thus blocking the holes. To remove this the sheets were hinged to make the reverse side accessible for occasional scraping and cleaning.

(5) Various hole sizes have been tried, but the best compromise has been found to be $\frac{3}{8}$ in. Trials with hard PVC sheet appear to be satisfactory.

The importance of good screening cannot be over-emphasised if one wishes to keep free from pump and drain blockage troubles.

PUMPS

The pumping of effluent is not without its troubles, and the choice of a suitable pump is important. It is desirable to have two pumping units; this facilitates cleaning and repairs.

The author has experience of an air flotation problem in the sumps which occurred due to foam formation lifting fat and fine hair to the surface, where it formed a 6-in-thick mat. This then broke up later and caused pump and blockage troubles. The problem was overcome by fixing a wooden distributor in place of a single weir, thus giving turbulence over the whole sump area.

SEGREGATION OF LIME AND CHROME LIQUORS

When considering effluent treatment one of the first decisions to be made is whether to separate certain of the more difficult liquors for special treatment or whether to treat the whole mixed effluent as one.

In my opinion the removal of sulphide from the lime liquors and the recovery of chromium from the waste tan liquors are better done by treating the liquors separately. This separation may be difficult to arrange in an existing tannery, but should be provided in a new tannery.

CHROME RECOVERY

Provided a tannery is of sufficient size—say about 4 million ft^2 a year—and if the waste chrome liquors contain not less than 0.4% Cr_2O_3, then at today's price it should pay to recover the chrome and re-dissolve it for further use. A method which can be used to achieve this is first to allow the waste liquors to stand so that suspended fleshy matter can sink. The clear liquor is then pumped off by a floating skim pipe and is treated with soda ash, the amount of which has been calculated from the analysis of the liquor. The precipitated chromium hydroxide is redissolved by direct addition of sulphuric acid and sodium sulphate to give a tanning liquor for re-use. Of course, there must be careful analytical control. The recovery of the precipitated chrome may be made with filter presses, rotary filters, or a centrifuge. My own preference, based on experience with all methods, is still the filter press, which can give a chrome paste containing 10% Cr_2O_3. A rotary filter gives only 8% Cr_2O_3 and a centrifuge even less. In the centrifuge the different particle sizes of the precipitate cause the paste to be a lumpy mixture, some of which dissolves easily but with a tendency for some lumps to be left undissolved.

The paste is dissolved with steam heating in a homogeneous, lead-lined, pressure vessel which can be self discharging provided the exit pipe dips to the bottom of the tank.

REMOVAL OF SULPHIDE

Lime liquors and certain of the first lime wash liquors should be segregated for treatment to remove the sulphide. Recovery of sulphide for re-use is not impossible but it is uneconomic. Since careful control, particularly of pH, is necessary it is much easier to do this on 6–8% of the effluent rather than on the whole effluent.

Several methods are known for the removal of sodium sulphide. All have disadvantages for different reasons.

The one which has probably been used most is treatment with ferrous sulphate with or without air blowing. The method which has been used in many textile dyeworks and at least one tannery in Great Britain is done on a batch process. The liquors are collected, the pH is adjusted to approximately 8·5 with acid, ferrous sulphate is added, followed by air blowing, which not only stirs the liquor, but also seems to have a coagulating effect on the ferrous sulphide precipitate. The precipitate is then removed by filter presses or settled out and removed in the usual way as a sludge. The filtrate goes with the rest of the effluent.

The disadvantages of this method are the careful control which is necessary—otherwise semi-colloidal FeS escapes and makes the effluent black—and the large increase in sludge produced; also the sludge containing FeS does not dewater and dry easily on the drying beds.*

* 1 ton sodium sulphide requires about $3\frac{1}{2}$ tons $FeSO_47H_2O$ and produces about 14 tons liquid sludge, 8% solids.

Treatment of lime liquors with chlorine gas will remove sulphides, the main reaction being:

$$Na_2S + Cl_2 = 2NaCl + S$$

Other side reactions occur to form oxysulphur acids which of course end up as the sodium or calcium salts. Much of the dissolved protein matter is also precipitated with the sulphur. Unfortunately, the cost of chlorine and the quantity required makes the treatment costs prohibitive.

Some time ago trials were made with flue gas which was passed up a tile-filled tower down which flowed the sulphide containing lime liquors. These were discontinued because of liquor filtration difficulties, scaling up in the reaction tower and because the SO_2 in the flue gas produced some colloidal sulphur on reaction with the soluble sulphide.

Another possibility of removing soluble sulphide consists of treatment with air in the presence of manganous salts which act as catalysts. The British Leather Manufacturers Research Association has developed the oxidation of sulphide by air using a manganese catalyst in a practical process. Up to 30 mg/litre of manganese are required. The air and effluent are brought into contact in a reaction tower by aerating with a diffuser, or by using mechanical aerators as in the activated-sludge process.

The removal of sulphides by the use of percolating filters has been practiced for some time and on a large scale in the rayon industry. Evidence is accumulating which indicates that if tannery effluent is eventually to be treated on percolating biological filters to remove dissolved protein matter, then the sulphides will be disposed of at the same time. This will mean, of course, that the filters will have to be large enough to cope with the extra load provided by the sulphide, and the maximum concentration of sulphide should not exceed 10 mg/litre.

FeSO₄ in Dorr Clariflocculator

Trials were made on a Dorr Clariflocculator plant to remove sulphide in the whole effluent, without segregation of the lime liquors, by pH adjustment and addition of copperas. With the existing equipment it was found difficult to exercise a fine enough control and at times the effluent was black with semi-colloidal FeS and the suspended solids figure seriously deteriorated. In addition, the sludge would not dry quickly enough on the drying beds, all of which became completely filled.

ACID v. ALKALI

It is interesting to look at the acid–alkali balance of a typical upper-leather tannery.

Alkalis (per annum)

$$\left.\begin{array}{l} 200 \text{ tons } Ca(OH)_2 \\ 95 \text{ tons } Na_2CO_3 \\ 38 \text{ tons } NaHCO_3 \\ 190 \text{ tons } 60\% \ Na_2S \end{array}\right\} = 433 \text{ tons NaOH}$$

Acids (per annum)

$$\left.\begin{array}{l} 236 \text{ tons } H_2SO_4 \\ 30 \text{ tons other acidic substances as } H_2SO_4 \end{array}\right\} = 220 \text{ tons NaOH}$$

The difference is 213 tons Equivalent NaOH.

This is sufficient to give a figure of 356 p.p.m. NaOH in all the effluent. Fortunately, some neutralisation is effected by carbonation, precipitation, etc., but it does illustrate the predominantly alkaline character of the effluent with which we have to contend.

REMOVING SUSPENDED SOLIDS

Originally, in a tannery in the north-west of England, two rectangular settling tanks in series were used to remove the suspended solids. As the production of leather increased over a few years, the total retention period dropped to as little as 2 h. In spite of this, these tanks produced about 30 tons weekly of solid sludge containing 14–18% solids.

In average weather on good cinder beds, the primary sludge would become solid and shovelable in about three days; the secondary tank sludge required twice as long. One tank had three baffles and was, as a result, less efficient. Contrary to a common belief, baffles increase the rate of movement and produce localised turbulence which does not assist settling.

Type of Plant

It was decided to replace the existing plant with a modern type self-desludging tank which unlike the original tank would require very little labour. After many preliminary tests and enquiries it was decided to ask one well-known firm of chemical engineers to investigate the properties of the effluent and design a plant.

Preliminary Laboratory Trials

During these trials an artificially mixed effluent and also many spot samples from the drain were placed in glass tubes 5 ft × 1½ in and observed for their settling behaviour. Since it was known that gentle turbulence for limited periods sometimes gives improved settling due to the larger particles absorbing the smaller ones, so the tests were repeated, but before standing the tubes were rolled slowly backwards and forwards across the bench. It was seen that this caused more rapid settling and smaller volumes of concentrated suspended matter at the bottom of the tubes. Thus aggregation was taking place and it was deduced that preflocculation followed by simple settlement should reduce the suspended solids from an average of 1000 mg to approximately 60–80 mg, provided the retention period was 24 h. There is a quiescent period of about 10 h during the night when there is practically no flow.

Variability of Effluent

One of the difficulties in treating tannery effluent is the ever-changing composition and flow rate. High flow-rate peaks are built up early in the morning particularly, when the worst polluting soak and lime liquors tend to be discharged. A long retention period helps to smooth these peaks out.

Details of Dorr plant

Balancing tanks are, in my opinion, unsatisfactory, because unless they are violently stirred, they become settling tanks which need desludging. The plant eventually chosen to replace the original rectangular settling tanks was a Dorr Clariflocculator of a size able to cope with 500 000 gal daily. The diameter was 102 ft and the central depth 17 ft. The sludge scraper rotates once in approximately 15 min. The inner flocculating compartment was 35 ft in diameter and fitted with 12 fixed paddles and 10 movable ones. Previous experience had shown the need for trapping fat which floats across the water surface, so a surface baffle ring and a descummer were fitted. The fatty matter was not attractive enough to sell so it was pumped back into the centre and became absorbed in the rest of the sludge.

Efficiency of Dorr Plant

To assess the efficiency of the plant, composite samples were made up by mixing hourly samples taken between 6 am and 6 pm. Analyses were made the following morning, the samples being kept in a refrigerator overnight. Even the daily mixed samples varied enormously, as these extremes over a four-week period show:

	Minimum	Maximum
pH	8·7	11·7
Total solids	3516	6050 mg/litre
Suspended solids	350	1756 mg/litre
Oxygen absorbed from N/80 $KMnO_4$ in:		
3 min	40	248 mg/litre
4 h	91	354 mg/litre
Sulphide (H_2S)	1	86 mg/litre
Sulphite (SO_2)	5	44 mg/litre

When the plant was run for a four-week period without pH control or addition of flocculants the following were obtained:

	Crude	Treated
pH	10·4	9·9
Total solids	4567	3781 mg/litre
Suspended solids	1128	148 mg/litre
Oxygen absorbed from N/80 $KMnO_4$ in:		
3 min	185	137 mg/litre
4 h	290	213 mg/litre
Sulphide (H_2S)	49	41 mg/litre
Sulphite (SO_2)	32	30 mg/litre

Removing Sludge from Dorr Plant

The desludging pump has a variable stroke and at the time of this trial drying beds were in use and the sludge was withdrawn slowly over about 4–5 h of the working day. On an average 65 tons containing approximately 18 % solids were obtained each week.

Later, in periods of wet weather or frost in winter, the drying beds were found to be inadequate, but it was found possible to dispose of the liquid sludge to a contractor who removed it by tanker. This liquid sludge averages about 7–8 % solids and even though it is of the consistency of thick porridge it is possible to pump it 350 yd to a loading bay through a 3-in pipe. Pressures vary between 25 and 40 lb/in². To accommodate the contractor sludge had to be removed rapidly in half-hour periods at 7.30 am, 11.00 am and 1.30 pm. Contrary to expectations, the quality of the effluent did not deteriorate as the following averages show:

Results with tanker

	Crude	Treated
pH	11·1	10·9
Total solids	3882	3577 mg/litre
Suspended solids	959	97 mg/litre
Oxygen absorbed from N/80 KMnO$_4$ in:		
3 min	147	132 mg/litre
4 h	259	211 mg/litre
Sulphide (H$_2$S)	44	41 mg/litre

Trials with pH Control

Trials were next made to ascertain the behaviour of the plant when the pH of the effluent was reduced by the addition of sulphuric acid. This was added manually at the head of a 200-yd open drain when it was known that large quantities of alkali were being discharged. Between 10 and 15 cwt were added daily, mostly during the morning, nearly four tons being used during the week of the trial. The aim was to enable the plant to operate at a pH of between 8 and 9. On the Thursday even 15 cwt of acid was insufficient.

Effect of Acid

The acid would: (a) neutralise lime; (b) neutralise Na$_2$S and enable some loss of H$_2$S to take place into the air; (c) precipitate some sulphur by interaction of sulphide and sulphite; (d) possibly precipitate some organic matter.

The results obtained (averages) were as follows:

	Crude	Treated
pH	11·4	8·65
Total solids	4515	3970 mg/litre
Suspended solids	925	87 mg/litre
Alkalinity to phenolphthalein (CaCO$_3$)	360	19 mg/litre
Oxygen absorbed from N/80 KMnO$_4$ in:		
3 min	196	106 mg/litre
4 h	304	137 mg/litre

Sulphide (H_2S)	79	12 mg/litre
Sulphite (SO_2)	29	33 mg/litre
Ammoniacal N	15	23 mg/litre
Organic N	92	24 mg/litre

Suspended solids were only slightly (if any) reduced. The $KMnO_4$ figures have dropped more than we dared to hope but the sulphide drop is excellent. Unfortunately, the addition of acid caused difficulties with the sludge. There was erratic behaviour during sludge pumping. Instead of a steady stream of porridge, the consistency varied from extreme watery thinness to abnormally thick lumpiness. This seems to be due to 'channelling' in the clariflocculator. One might have expected the sludge to dewater easily but, on the contrary, it did not dry well on the beds.

Trials with $FeSO_4$ and aluminoferric did not offer any advantage, but a precise assessment of these flocculants needs a more accurate pH control and dosing equipment.

Summarising, the Dorr Clariflocculator is capable of removing just over 90% of the suspended matter in a grossly polluting tannery effluent when the pH is kept in the region of 8 to 10. The permanganate figure is reduced to approximately 50% of the original and a substantial removal of sulphide takes place.

BIOLOGICAL TREATMENT

Obviously tannery effluent requires final biological treatment to remove the dissolved organic matter if complete purification is required. It is fairly well established that quantities of tannery effluent up to a load of 30% can be mixed with domestic sewage and passed through the town plant without too much difficulty. The normal percolating filters seem to cope quite easily with such mixtures of sewage and effluent. The activated-sludge process is rather more easily disorganised by flushes of obnoxious chemicals and takes much longer (perhaps weeks) to recover.

Vegetable tannins, however, do impart a brown colour to treated sewage though the appearance may be worse than the analytical figures indicate.

Biological Filters

The behaviour of biological percolating filters when fed with neat tannery effluent is less understood. Since there will be considerable differences between the effluents from different tanneries producing different types of leather, considerable pilot-plant work at the tannery will be necessary to see how much purification is obtainable. Certainly, the pH must be controlled and the flow balanced. It is known that certain tannery effluents are treatable on percolating filters and that any sulphide is oxidised away at the same time. Opinions differ as to the effect of small amounts of chrome salts on biological oxidation. Recent evidence suggests that the process is not so easily upset as some sewage works managers believe. Chrome to many people immediately suggests bichromate and some managers may be confusing the issue. Trivalent chromium is used in tanning.

The activated-sludge process is unlikely to be suitable for undiluted tannery effluent. It is too easily upset by many chemical products used in the tannery and it requires a degree of control more suitable for the sewage-works chemist than as a part-time job for the tannery chemist.

DEALING WITH SLUDGE

One square foot of chrome upper-leather will produce during its manufacture not less than 1 oz of solid suspended matter. From an efficient settling plant, this will be equivalent to about $1\frac{1}{2}$ lb of liquid sludge (8–9 % solids) or about half the amount of solid sludge. If unlimited land area is available (which is rare), then the cheapest method of disposal is to pump it on to lagoons and allow it to dry over a prolonged period.

If a suitable dumping space such as a disused quarry is available within about 6–7 miles, then there appears to be a marginal advantage in removing the liquid sludge off the premises by tanker, rather than drying it on beds and then removing it as a solid.

Under present conditions of labour shortage, people are less willing to accept the heavy work involved and the unpleasant nature of sludge. Also, dumps of solid sludge in towns breed vermin and create odours, so removal as liquid sludge is to be preferred if at all possible.

Sludge Digestion

Some time ago it was decided to enlist the help of the Water Pollution Research Laboratory to investigate the anaerobic digestion of tannery sludge. Preliminary experiments using the Warburg respirometer at 33°C, over four days in an atmosphere of nitrogen, showed that tannery sludge alone will digest and liberate gas. A mixture of the chrome tannery sludge with sewage sludge digested at a faster rate.

The work was then transferred to 3-litre digesters equipped with stirrers. Tannery sludge alone and a 1:1 mixture of tannery and sewage sludge were digested at 35°C and 45°C. The daily volume added was stabilised to a 25-day retention time and the daily load was 0·03–0·04 lb organic carbon/ft^3 digester space.

Average results during 8 weeks 'steady state' operation
(Retention time 25 days)

Influent sludge	Temp. of digestion (°C)	Daily carbon load (g)	Volume gas (ml/g carbon load)	Conversion organic carbon to gas (%)	Volume digested sludge settling in 24 h (%)
Sewage	35	1·59	963	48·6	69
	45		956	48·3	89
Sewage/ tannery 1:1	35	1·84	800	40·4	63
	45		718	36·3	83
Tannery	35	2·09	480	24·2	74
	45		212	10·7	92

The average composition of the sludge used in all these trials was as follows:

	per cent
Total carbon	1·95
Organic carbon	1·52
Total nitrogen	0·26
Free NH$_3$	0·03
Total solids	7·28
Suspended solids	6·80

The next stage of the investigation was carried out over a period of two months using a 15-litre digester. Sludge was fed in once daily and the temperature was kept at 33°C. The return of settled sludge to the digester was investigated and the retention times were varied.

The results of these trials may be summarised as follows:

Chrome upper tannery sludge can be digested anaerobically with the removal of 42% of the organic carbon when the daily load is 0·03–0·05 lb/ft^3 digester space at a temperature of 33°C. The retention time was 21 days. 85% of the carbon removed was evolved as a gas containing 70% of methane and 30% of CO_2. The efficiency is somewhat less than sewage sludge and much less than slaughterhouse and distillery sludges.

One ton of sludge per week requires approximately 110 ft^3 of digester space and the quantity of gas evolved would provide sufficient heat to maintain the digester at 33°C.

Further larger scale trials are necessary to establish the detailed economics of the process. The reduction in volume (20–25%) is welcome though not spectacular, but the digestion made the sludge less amenable to drainage and dewatering if it had to be dried on cinder beds or filtered.

It is interesting to note that nothing in the tannery sludge inhibited the digestion and no trouble was encountered with the production of H_2S.

This work would seem to prove that if tannery effluent was discharged into the local authority sewerage system it would have no detrimental effect on the operation of the domestic sewage digestion plant.

CONCLUSION

It is now possible to see why this chapter has been called 'The Tannery Effluent Problem'. It is really an economic problem. It is certainly possible, using existing knowledge and techniques, with perhaps some modifications, to substantially purify effluent. The important questions are: To what standard and at what cost?

To achieve Royal Commission Standard would mean that a medium-sized tannery employing say 300–400 people would have to finance and maintain a sewage works not less in size and probably more complicated than that operated by a town of 60–70 000 people. The capital cost alone may well be not less than a quarter of the cost of the tannery and its equipment. This alone would put more than a penny per square foot on the leather before the costs of running the plant have been taken into account. The costs are not likely to be much different if the effluent is handed over to a local authority. At the moment, there seems to be little likelihood of any revolutionary cheap process of treatment coming along.

Clearly, governments, local authorities, and river authorities will have to decide whether a leather industry is necessary for the well-being of a country or locality. If the answer is that tanneries are necessary, then some degree of environmental pollution may have to be accepted. Geographical location, climate, and national economics are of paramount importance in deciding this issue.

Chapter 11

TREATMENT OF WASTES FROM TEXTILE BLEACH AND DYE WORKS

A. H. LITTLE

Variability in composition is a marked feature of most textile works effluents and wide variations in composition, temperature, flow, and pH value are prevalent. It is essential to blend in the concentrated wastes with dilute wash waters and to neutralise strong acids or alkalies so that substantial uniformity is obtained in the material presented for treatment. The wastes differ according to the fibres being processed, but the main steps in treatment are sedimentation to remove solid matter and biochemical oxidation to remove soluble organic matter. Textile wastes are readily oxidised in either the percolating filter or the activated-sludge processes, although there is often a deficiency in nitrogen that has to be made up. The pollution load can sometimes be reduced considerably by making changes in the chemicals used or by alterations in the textile processes employed.

CHARACTER OF TEXTILE WASTES

In the bleaching and dyeing of textile materials there may be some continuous processes, but there are many batch treatments, such as kier boiling and winch or jig dyeing, that give sudden large changes in composition as the machines are emptied. Temperature and pH will also change abruptly, and of course the flow alters with each surge of waste. The different fibres also affect the composition of the waste; cotton and linen contribute much organic matter from the non-cellulosic materials that are present in the natural fibres, while wool contains sand, grease, and suint which are removed in scouring. Cotton may contain from 6 to 10% of non-fibrous matter, whereas wool may contain from 20 to 70%, most of which may be removed in scouring or other wet processes. The synthetic fibres are relatively free from additives, with only added spinning oil and anti-static dressing, but the quantities are not negligible. The composition of a works effluent will therefore vary with the fibres that are in process. Apart from natural impurities and added dressings most fabrics woven from both natural and synthetic materials will contain large amounts of sizes of various kinds used to protect the yarns in weaving, and the removal of these adds a substantial load to the effluent. The amount of organic matter that is removed from a fabric in the course of normal textile processing can be visualised when one considers that about 10% of the gross weight of a cotton fabric consists of natural impurities and size, and may be removed in processing. For a firm that handles 20 tons of fabric a week, a not unusual scale of working, this means that 2 tons of impurities go down the drain and have to be removed from the effluent.

PRELIMINARY SURVEY OF WORKS

At the start of an investigation to select a suitable purification process for a works effluent, one is faced with the problem of finding the composition and variation of a very complex and very variable mixture of substances. It is helpful to locate the main sources of organic matter and of strong alkali or acid residues, and to determine the times and volumes of such discharges. This can be done by selective measurement of machine volumes and times of emptying and sample collection, from which can be worked out the contributions to the main flow of the various machines and processes. Variations in flow rate, pH, temperature, and organic load (BOD) through the working day should be known and these will give a fairly good indication of the scale of working that will be required. Usually they also indicate the need for balancing of flow and for neutralisation, as it would be difficult to treat wastes that are as variable as most textile effluents are.

To balance the flow, temperature, and composition it is necessary to resort to deliberate mixing on a large scale. The discharge from a bleach or dyeworks changes over the day according to the processes that are running, and one must mix on such a scale that strong liquors from one period are mixed with weak wash waters from another. The extent to which this has to be done can be brought down if some of the strong wastes are segregated and stored separately, letting them off into the main effluent stream continuously over the day, so that they are well diluted. This is a good method of dealing with such contaminants as kier liquors or sulphur dye baths.

In the course of a survey by sampling, the range of composition can be estimated and, from this, the amounts calculated of solids and soluble organic matter that have to be removed. These figures are essential to the designer in working out the size of plant needed to purify the waste, but in practice no very precise figure is possible from sampling alone because of the intrinsic variability of textile wastes.

Before or during mixing is the appropriate time to neutralise excess alkali or acid. Many textile wastes are alkaline, some very strongly alkaline, and addition of acid, controlled by a pH meter, is effective in removing the excessive alkalinity. The total amount of acid needed can be estimated fairly accurately from the figures for alkalies and acids used in the works over a period of six to twelve months. A balance can be struck and any deficiency in acid has to be made up by adding sulphuric acid to the effluent. On strongly alkaline wastes such as those from mercerising machines, the amount of acid needed is very great, so that neutralisation can be expensive if mineral acid is employed. A somewhat cheaper method is to use as the acid carbon dioxide obtained from burning propane or fuel oil under the liquor surface by so-called submerged combustion. This process can be regulated by a pH meter and it requires little in the way of equipment. Still cheaper is the use of carbon dioxide from flue gas, the alkaline liquor being treated with excess of the flue gas in a simple absorption tower. The size of tower can be calculated from the amount of alkaline waste to be treated and the carbon dioxide content of the flue gas. The main difficulty with this method is in overcoming the prejudice of the works engineer in diverting some of his flue gas to process use. However, several plants using flue gas are now in use, so some conversions have been made.

SEDIMENTATION

In textile trade wastes the solids in suspension come from the natural fibres, from materials used in sizing warp yarns for weaving and from residues of dyes or finishing materials. The amounts present depend upon many factors but, as with soluble impurities, tend to be highest with woven fabrics made from natural fibres and lowest with knitted fabrics made from synthetics. Low values may also be found in wastes from package dyeing of yarns where the packages themselves make excellent filters and remove most of the solid particles from the dye liquors.

Suspended matter is normally removed by a process of continuous sedimentation, in large tanks where the big volumes of bleach and dye liquors flow through slowly and remain for a sufficient time for the solid matter to sink to the bottom as a sludge. This is fine in theory, but in practice the picture is very different. The liquids do not flow evenly through the tanks; the momentum of the entering liquid sets up currents and eddies in the tank, and where differences in temperature and density occur, as is often the case with textile wastes, the liquids flow to various levels; the dense ones to the bottom and the light ones to the surface. This stratification makes the effective size of the tank much smaller and the lighter (or hotter) liquids flow through preferentially. Consequently, the efficiency of removal of suspended matter is often very poor, especially with the fine particles that come from textiles, and settling tanks may not remove anything like the amount of solids that they are expected to.

Some improvement can be obtained by the use of flocculating agents such as alum or iron salts, which result in larger particles for sedimentation and consequently more rapid fall to the bottom of the tank. With such agents control of conditions of treatment has to be strict. In alum flocculation the rate of floc formation is at an optimum in a narrow pH band near the neutral point, and outside this range it is less effective.

Flocculating agents also increase the amount of sludge that has to be removed and the disposal of this sludge is always a troublesome matter. On the other hand, apart from the assistance in removal of solids, the use of flocculating agents in some circumstances may be an advantage as the flocs absorb some of the dissolved organic matter and reduce the quantity that can cause river pollution. Unfortunately, with textile wastes the amount removed is usually only about 20 to 30% of that present, so the treatment cannot be relied upon to produce a major improvement in the wastes. The method is effective, however, in particular cases, such as the removal of residual grease from wool scouring liquors. In the treatment of those liquors substantial amounts of ferrous sulphate and lime are used, controlling the pH near to the neutral point, and the flocculated mixture of ferrous hydroxide and grease is removed by sedimentation, preferably by static settling, or by filtration.

Reverting to consideration of the behaviour of sedimentation tanks it is worth while mentioning that the amount of sludge that has to be removed from a tank is a good measure of its efficiency. Many tanks are thought to be efficient, but on enquiry it is often found that they are desludged once a year, at the annual shutdown. A simple calculation from the amount of waste flowing through and the content of suspended matter shows how much solid matter should be there, and if this indicates that the tank should be desludged once a month, or even at shorter intervals, the inefficiency of settlement is made obvious.

For small volumes of effluent it would pay to have two tanks each holding the flow for 24 h and run these on the 'fill and draw' principle, one filling over a day and the other settling under static conditions, with the clear liquid drawn off towards the end of the period so that the tank is ready for refilling the next day. Static settling avoids many of the difficulties of continuous flow tanks and gives better results, but requires a little more attention.

BIOCHEMICAL OXIDATION

Much of the soluble organic matter in the waste liquors can be removed by treatment in the presence of micro-organisms, under conditions of good aeration. In a suitable environment the organisms absorb the organic matter and use it and the dissolved oxygen in the water to live and grow. When sufficient organisms are present and the conditions of aeration, temperature, pH and nutriment are correct the growth of the organisms and the absorption of the organic matter are astonishingly rapid. Large amounts of bio-degradable organic matter can be removed leaving a liquid either substantially free from organic matter or containing resistant material that the organisms cannot break down or cannot tolerate. In practice it is found that starting with a mixed population the organisms thrive that can deal with the components of the effluent, and with time the efficiency of removal increases until it reaches a maximum after several weeks or possibly longer. This again shows the need for balancing the composition of the wastes, for if marked fluctuations take place, the optimum growth cannot be achieved.

In addition to the soluble organic matter, the organisms need small amounts of nitrogenous material, either organic or inorganic, and phosphorus-containing matter such as phosphates. These are essential to the cell growth and if there is a deficiency they have to be added. Usually there is enough phosphate in textile wastes, but with synthetic fibre wastes there may be a shortage of nitrogen, and ammonium sulphate has to be added to make up for the deficiency.

In addition to removal of organic matter in the biochemical process it is found that sulphides are removed almost completely if the initial concentration is not too high. Sulphur dye liquors can therefore be dealt with provided they are mixed and diluted in the main effluent stream.

It is necessary to avoid the inflow of substances to the waste that could upset the microbiological metabolism. Such substances that might be in textile wastes are chromates, copper salts, rot-proofing agents such as lauryl-pentachlorphenol and other chlorinated materials. These would slow down or bring to a halt purification processes, and must be kept out.

Two methods of biochemical treatment have been employed for textile wastes, both adapted from use in sewage purification. The first is the 'percolating filter' in which the organisms grow on a wetted solid medium exposed to air. The second is the 'activated-sludge' process in which the organisms grow in the liquid phase, in the presence of much oxygen from intense aeration of the liquid.

Generally, a percolating filter is more expensive to install than an activated-sludge plant, but requires less attention in use. An activated-sludge plant may cost only half that of the other type although it may need close attention if something goes wrong with the organisms, resulting in ineffective removal of organic matter. Care has to be taken also in maintaining the high concentration of organisms by returning to the system sludge from the subsequent sedimentation, as the efficiency is roughly proportional to the concentration of organisms present in the aeration tank.

PERCOLATING FILTER
This method of purification applied to textile wastes has indicated that the organic matter in textile wastes can be removed readily once the system has become acclimatised to the waste.

In trials starting with a new filter, without any inoculation with micro-organisms, the requisite population built up in three to four weeks and removal of organic matter was rapid in one pass through the system.

When the system contains an active population the throughput rate can be increased, but at high rates the removal efficiency starts to fall off. To achieve high rates and at the same time produce a well-purified effluent it may be necessary to run two systems in series, a roughing filter followed by a finishing filter. This gives excellent removal of organic matter but is expensive in equipment.

The plastic medium 'Flocor' (ICI) is intended for use as a roughing filter on strong wastes, but it has been found effective on dyehouse wastes of moderate strength giving effective removal of organic matter at rates higher than are normally obtained with stone or slag filters. This may allow an advantage on cost. Although the medium is comparatively expensive less would be needed and, as only a light enclosing structure is required, the expenditure on civil engineering work is much lower than for conventional filters.

ACTIVATED SLUDGE

Treatment is in tanks in which the liquor may be aerated in a variety of ways. Compressed air passing through diffusers can give fine bubbles, or rotating agitators can beat the surface, creating turbulence and entrainment of air. The objective is to produce great turbulence with many air bubbles, so that mixing and dissolution of oxygen can proceed rapidly. With a high concentration of active micro-organisms the oxygen in solution is used up rapidly and quick replenishment is needed. At normal temperature the solubility of oxygen is near 9 p.p.m. but in an active system 2 p.p.m. would indicate efficient aeration while 1 p.p.m. or less is not uncommon. At the lower rates the reactions still proceed, but are limited by the amount of oxygen available.

Laboratory trials on cotton dyehouse wastes show that they are readily purified by the activated-sludge method but in some cases the biological system builds up only slowly. Wool scouring wastes on the other hand give a rapid build-up of organisms. The difficulty with the cotton dyehouse effluents is partly due to lack of nitrogenous material, but even when ammonium salts are added the build up is comparatively slow. Respirometer studies have shown that the absorption of oxygen by carbohydrates in an inoculated system is slower when nitrogen is added as ammonium sulphate than when added as glutamic acid, and this seems to indicate that the availability of the nitrogen as well as the quantity may be of importance.

A sufficient quantity of sludge present in the aeration system is essential as this sludge contains the active components and their concentration is an approximate indication of the capacity of the system. It is therefore necessary to return any organisms that pass out of the aeration tank. These are normally recovered as a sludge from a sedimentation tank and passed rapidly back into the system. With textile effluents the sludges have been found to settle readily and so far no difficulty has been encountered in separating the sludge and returning it to the aeration tank.

SEDIMENTATION AFTER BIOCHEMICAL TREATMENT

The liquor coming from a percolating filter or from an activated-sludge plant, particularly the latter, contains considerable amounts of solid matter composed mainly of organisms that have been taking part in the oxidation. These solids are mainly in the form of small flocs that can be removed by sedimentation, and the clarified liquid passed on. In the activated-sludge process the solids are returned quickly to the aeration tanks, in order to maintain the high concentration of organisms that is essential to this method of treatment. In handling the returned sludge it has been found desirable to avoid violent turbulence as this disperses the floc. Screw or ram pumps appear preferable to centrifugal pumps. Any dispersal of floc in the returned sludge may show in high suspended solids on the final outflow, because the fine particles being slow to settle may not be removed in the normal time of sedimentation.

REDUCTION IN LOAD ON THE TREATMENT PLANT

The size of plant required for treating the waste from a textile works will depend upon the organic load involved. The flow and the amount of suspended solids will set the size of sedimentation tanks and the organic matter in solution will govern the size of the aeration tanks needed. If the organic load can be lightened then a smaller plant will be required or the load on an existing plant can be reduced. Sometimes simple process changes will bring down the load. The BOD of chemicals used gives a guide to how these substances will behave in the effluent treatment process, and materials with a high BOD can be avoided in favour of others with a low BOD. Thus the use of formic acid instead of acetic acid gives a reduction in BOD of about two-thirds, while the use of a mineral acid would eliminate the source of BOD altogether. Soaps have high BOD levels whereas non-ionic detergents are generally low. With detergents, however, the choice of objective is important. If removal of surface-active matter is the target a 'soft' detergent with a high BOD is unavoidable.

On the other hand removal of organic matter, as shown by the BOD, can be achieved by the use of a 'hard' detergent with a low BOD. Starch with a high BOD might be replaced by carboxymethyl cellulose and give a lowering of BOD in the ratio of 50:3. This might be of value to vertical organisations where the sizing materials can be selected according to their behaviour in wet processing and their contribution to the effluent organic load. It is not helpful to those firms that have to accept fabric woven elsewhere, in which the finisher has little or no say in the materials in the cloth.

Wastage of chemicals can be kept down by good housekeeping within the works and when spillage of materials can be avoided there is less washed down the drain to increase the load on the treatment plant.

In the actual processing a choice of method may often have considerable effect on the effluent, particularly in the field of cloth preparation. Thus the traditional practice of kier boiling may give kier liquors with very high BOD levels in the region of 10 000 p.p.m. or more. A peroxide bleach in the kier will give a much lower BOD level, usually less than half that of a kier liquor, while a chlorite bleach liquor would be still lower, possibly only a tenth of the kier liquor value. As the kier liquors in bleaching may represent half the BOD load in the effluent, the substitution of one of the other processes may effect a dramatic reduction in the quantity of organic matter in the waste liquors. In any consideration of process changes it is well worth while to bear in mind the ultimate disposal of any process liquors used. It is fortunate that present-day trends in bleaching techniques are towards processes that reduce the load on effluent treatment plant.

BIBLIOGRAPHY

'Effluent Treatment and Water Conservation', Final Report of the Effluent and Water Conservation Sub-Committee of the Committee of Directors of Textile Research Associations, Shirley Institute, Manchester, 1968.

Little, A. H., Treatment of textile waste liquors, *J. Soc. Dyers & Col.* (1967), **83**, 1967.

Little, A. H., Treatment and control of bleaching and dyeing wastes, *Water Pollution Control* (1969), **68**, 178.

Chapter 12

TREATMENT OF PAPER AND BOARD MILL EFFLUENTS

T. WALDEMEYER

It is a truism to say that papermill effluent problems start at the water intake to the mill and that, before methods of effluent treatment can be selected, it is essential to study the whole water economy of the processes used in papermaking. Although it does not generally appear in the final product, water is one of the main raw materials used in the manufacture of pulp and paper. Compared with many other industries the water consumption per ton of product is very high, 20 000 gal of water per ton of paper being an average figure for Great Britain. Most of this water reappears as effluent and the paper industry has had to face some very difficult effluent problems in order to reduce pollution of streams and rivers.

This chapter will only deal incidentally with the very large volumes of strong effluents that are inevitably produced in the preparation of pulp from the forests of North America or Scandinavia. The final yield of pulp may only be 50% of the original tree substance and, apart from the solid wastes such as barks, most of the other material leaves the mill in the form of suspended solids and dissolved substances such as lignin derivatives, gums, sugars, and other soluble salts. In the (acid) sulphite pulp processes some of the strong liquors produced may have BOD's up to 35 000–50 000 p.p.m., and the total BOD load from the pulp mill may be 550–750 lb BOD/ton of pulp.

For comparison a 'neutral sulphite' semichemical pulping process which can be used for hardwoods, to give a 70–80% yield, may produce a BOD load of 140 lb/ton of pulp. This may be reduced if modern methods of chemical recovery are incorporated in the design of the pulp mill. In the case of Kraft or sulphate pulp, efficient chemical recovery systems are an integral part of the process and the BOD effluent load is much lower, say 50 lb/ton. As many mills may produce over 500 tons of pulp each day the sheer size of the problem can be appreciated. In England one pulp mill is sited on the Severn estuary and a large new one is in Scotland. Some mills in Scotland are based on the use of esparto grass as a raw material; some straw is also used in England, together with small quantities of rags and other materials.

Although a figure of 20 000 gal of water was quoted above as the average water consumption per ton of paper produced, this average covers a range of from 5000 to over 100 000 gal/ton and each mill presents an individual problem depending on its geographical situation and the availability of water, the age of the paper machines, the types of paper which are made and the nature of its production, *i.e.*, long or short runs. Modern paper or board machines are complex and run continuously at high speed for six or six and a half days a week, the shutdown being followed by a complete washing out of the system; this creates special problems, especially in large mills which may have fifteen machines. In the design of new machines it is now customary to give special attention to the design of the water circulation systems so as to minimise losses, and each machine is provided with its own 'saveall' or fibre recovery system. These specialised pieces of equipment are based on sedimentation, flotation or filtration processes and they also serve to conserve water and other valuable materials such as fibre, clay, dyes, alum, and size. Heat is also saved and, indeed, the system may be so 'closed up' that water may have to be added to maintain adequate temperature control. Further sedimentation water and effluent recovery plants may be installed outside the mill to serve a group of machines with selective re-use of the settled solids. The recovered 'water' is used to supplement existing sources of water or to meet the need of expansion. Colour problems can arise at this stage and recirculation of the water within the mill can lead to a build up of soluble salts, ideal conditions for encouraging the growth of 'slime' organisms in pipes, storage chests, and odd corners of the system. When these break away they cause trouble on the machine, and it is nowadays routine practice to add bactericidal chemicals at the beater stage of the process. This factor must be borne in mind when considering biological methods of effluent disposal.

REMOVAL OF SOLIDS

In spite of the substantial amounts of fibre which are now saved in the papermaking system, by use of savealls, there are losses to the drains which cannot be prevented. Here the fibre comes into contact with floor washings, dirt and oil which makes the fibre unacceptable to the papermakers even if it were all recovered. Before solids can be removed from the general mill effluent it is often necessary to consider extensive redrainage of the mill in order to collect all the effluent at one point. This can be an expensive part of any effluent scheme, especially in old mills where there may be many existing outlets to the stream. At mills using imported pulp for their main papermaking operations the effluent problem is concerned only with the efficient removal of the suspended solids.

One common method of settling papermill effluents uses large lagoons. These are shallow excavations surrounded by earthen banks—the total depth varying from 3 ft to about 10 ft—with an inlet pipe at one end and an overflow drain or river at the other end.

They are cheap to construct and give good results if adequate capacity is provided. Removal of settled sludge is expensive and is rarely done often enough; lagoons can create a severe odour nuisance and occupy a large area of land.

Modern types of concrete sedimentation tanks, usually circular in shape, have been installed at quite a few papermills and they are giving good service. Diameters vary from 30 to 100 ft depending on the flow and on the nature of the solids in the mill effluent. Normal scraper gear removes the settled solids to a sump in the centre of the tank. These are discharged at a controlled rate; the concentration can be as high as 8–10% solids although it is more usual to run at about 3–4%. With finely divided clay present it is difficult to obtain an effluent with a suspended-solids figure below 30 p.p.m., unless the capacity provided is very large—say, using upward-flow rates of 1–2 ft/h. With design rates of 3–4 ft/h, the suspended matter in the effluent may average 50–60 p.p.m., of which over half may be mineral matter.

Settlement may be improved by adding flocculating chemicals such as alum or activated silica. The last named has given very good results at several mills where the existing sedimentation tanks were overloaded. Without increasing the size of the plant the suspended solids in the effluent have been reduced to an acceptable value. Generally, the use of chemicals is to be avoided as it greatly magnifies the sludge problems, both in increasing the quantity and in making it more difficult to dewater. More efficient settling may also be obtained by the use of physical 'flocculation' in the form of slow stirring by moving paddles. In the clariflocculator the central part of the circular clarifier is fitted with revolving vanes. It is claimed that the combined tank will be more efficient than the same size of normal clarifier, but a good deal will depend on the nature of the solids in each individual effluent.

Sedimentation is the normal way of dealing with papermill effluents, the use of the various rotary and vacuum filters being restricted to serve all systems within the mill or for dealing with the settled solids at a later stage. However, it is sometimes possible to segregate substantial volumes of the effluent containing relatively little suspended solids, say 50–70 p.p.m. Such an effluent arises from felt-washing operations and can be successfully treated by the use of microstrainers or rapid sand filters.

CONCENTRATION OF SOLIDS

Disposal of settled solids, as in many other industries, remains the main problem to be solved. The ideal must be to re-use the recovered fibre and other solids within the mill itself to make possibly a lower-grade paper—but there are very real difficulties involved as the papermaker is always striving for cleanliness and avoidance of any impurity or change of colour which could interrupt production on the fast-running machines.

If the recovered solids cannot be pumped directly to the point of use or if they have to be stored or transported elsewhere it is necessary to use some form of concentration process. Sludge-presses are used at some mills. These are of standard construction and produce a cake 2–3 in thick containing 30–40% solids. The feed sludge is drawn from sedimentation tanks at a probable average consistency of 3–4%. Heavy twill cloths are used. The cake is easily stacked and can be transported by truck to the beaters for direct re-use or by road to other mills. Labour costs with the old type of press are heavy.

Vacuum rotary filters are also used. Rates of filtration vary with the nature of the suspended solids, but the output would probably be about 3 lb of dry solids per square foot of filter per hour. The moisture content of the cake may be higher than for pressed cake obtained from the same material, say 25–30% solids. The use of a

'Rotoplug' machine may be economically justified where the quantity of solids to be handled is not too large. A continuously running type of centrifuge has been used both for the concentration of settled papermill and boardmill sludges. This gives a product which is dry enough to remove by lorry, but the liquor which passes through the centrifuge is heavily charged with finely divided suspended solids and must receive further treatment. Where suitable land is available disposal by discharge to storage lagoons and removal to tip by excavator and lorry is probably far cheaper than any mechanical method of dewatering, say under £4.00 per ton of dry solids. A large mill may have between 10 and 20 dry tons of solids for disposal every day if all the effluents from the mills are adequately settled.

REMOVAL OF ORGANIC MATTER

Methods available to reduce the BOD of papermill effluents are mainly based on the principles of biological purification processes already established for the treatment of domestic sewage and for many other types of industrial wastes. Aerobic processes such as percolating filters and activated sludge are used at present.

Where imported pulp is used as the main raw material for papermaking the BOD of the settled effluent will not be very high—between 100 to 200 p.p.m. would be an average figure—and the problems arise mainly from the large volumes of effluent which must be purified to meet the standards required by the River Authority. With the increasing use of auxiliary chemicals and other products in the papermaking process to meet customer's quality requirements there is a tendency for the general BOD of papermill effluents to increase. Boardmills and papermills making certain grades of packaging are using recovered waste materials as their main raw material for paper and board-making, e.g. the various grades of wastepaper. Owing to solution of ink material, glue, size, etc., during the reconversion process the effluent from such a mill can be very strong, especially if efforts are made to reduce the water consumption by recirculation. The total solids of the effluent may build up to a concentration of over 8000 p.p.m. with correspondingly high BOD, over 500 p.p.m.

Many mills abroad and some in this country also use a de-inking process to convert wastepaper into higher grades of pulp and effluents from such processes, although small in volume, will also have higher BOD's than the normal papermill effluent.

PERCOLATING FILTERS

Some mills situated on the upper reaches of clean rivers or at points upstream of domestic-water intakes have been using standard percolating filters for many years. One such mill has seven 100-ft diameter filters, with clinker medium, and the effluent after passing through a large lagoon area is of Royal Commission standard: BOD below 20 p.p.m., suspended matter below 30 p.p.m. The rate of treatment in this case has been high and, due to inadequate primary settling tanks, two of the old filters have become blocked with fibrous and inorganic matter and the medium has had to be renewed. In other cases rates of treatment have been similar to those used for domestic sewage, *i.e.*, loadings of 0·1–0·2 lb BOD/yd^3/day.

Southgate (1945) has reported experiments on the treatment of washings from digested straw by the process of alternating double filtration, the process first used successfully for the treatment of milk-washing wastes. He found that the wastes, when mixed with the available domestic sewage, could be treated at the rate of 0·07 lb BOD/yd^3/day giving a 69% reduction in BOD. Higher loading could be applied but only at the expense of a lower quality effluent.

A later paper from the Water Pollution Research Laboratory (Eden *et al.*, 1952) gave results for the treatment of wastes from a mill using hemp, linen, and cotton as raw material. The 'black liquor' from the cooking process had a BOD of 8600 p.p.m., permanganate value of 6800 p.p.m., and total solids 35 000 p.p.m. After dilution to give a feed with BOD 200 p.p.m. and addition of nutrients (P, K and N) it was found possible to purify this on a filter at rates of 0·13 lb BOD/yd^3/day to give an effluent with BOD 15 p.p.m. The general wastes were also found to be amenable to treatment although the high pH and residual chlorine might cause some difficulty. Efficient humus tanks are required to remove suspended solids from the filter effluent and recirculation of this settled effluent to mix with the incoming waste may allow higher BOD loadings to be applied to the filter—up to 0·25 lb BOD/yd^3/day—whilst still maintaining the same standard of final effluent quality. Recirculation also helps to reduce 'ponding' of filters.

Papermill wastes are usually deficient in nitrogen and the addition of suitable nutrients is often recommended to promote maximum biological oxidation. Experiments on a plant scale are by no means conclusive, and provided some domestic sewage is present in the waste, the expense of adding nutrients may not be justified. These remarks do not apply to the purification of pulpmill effluents.

Large-scale filter installations using synthetic plastic filter media have been used in the USA and at papermills in England. When used as roughing filters to give partial purification, BOD loadings of up to 4·0 lb BOD/yd^3 can be applied.

An article in the technical press has described a biological filter plant at a papermill in England which uses filter presses to concentrate the secondary humus sludges.

ACTIVATED SLUDGE

The standard activated-sludge processes have been used on a very large scale in the USA and several purification plants have been in operation in England for a number of years. The following section describes the activated-sludge plants in use at mills of the Reed Paper Group of which the writer has had personal experience.

The activated-sludge process of purification has been used to treat the effluents from the Tovil and Bridge Mills before discharge into the River Medway and at Reed Board Mills (Colthrop) to purify papermill and board mill effluents before discharge to the River Kennet.

The effluent treatment plant at Bridge was put into operation in 1957. Briefly the mill effluent from two machines making straw-paper, and some excess backwater is settled in a standard concrete clarifier (100 000 gallons), the recovered fibre being re-used within the mill. The effluent is then pumped to a storage lagoon (400 000 gallons capacity) and drawn off under controlled conditions (3000 gal/h) to the inlet of the aeration plant and mixed with an equal amount of surplus process water. The small amount of domestic sewage from the mill is also introduced to provide a continuous supply of active bacteria and act as a source of nutrients. No chemicals are added as nutrients and although the wastes are probably deficient in nitrogen in relation to the BOD load entering the plant, there is no definite evidence that the addition of ammonia or phosphate could increase significantly the throughput of waste to the plant and still maintain a quality of effluent below Royal Commission standard: BOD 20 p.p.m. and suspended solids 30 p.p.m. The aeration tank capacity is of 53 000 gal capacity giving

T. WALDEMEYER

TABLE 12.1
Bridge Mill activated-sludge plant
Results in parts per million

Period	Waste					Effluent					Mixed liquor	
	TS	SS	pH	PV	BOD	TS	SS	pH	PV	BOD	SS	% Ash
Average: April–October 1960	1 440	77	7·7	130	227	780	22	7·6	21	4·9	2 980	50·6
Average: November 1960–March 1961	1 440	150	7·9	106	200	780	39	7·6	25	11·7	2 310	51·3
9 September 1963[a]	1 200	150	7·9	58	180	720	Tr.	8·0	8·2	3·8		
4 November 1963[a]	1 650	400	7·4	31	295	700	34	7·9	9·6	4·6		
13 January 1964[a]	1 500	290	6·6	190	750	850	18	7·8	8·0	3·0		

TS: Total Solids.
SS: Suspended Solids.
PV: Permanganate Value (4 hours 'oxygen absorbed' test).

[a] Samples taken and analytical results from W. F. Lester, then Chief Inspector, Kent River Board. Corresponding analyses for the dilution water entering the plant show 6·5 to 10·0 Nitrate N and 2·0–4·8 p.p.m. BOD.

12 to 18 h aeration (9 h calculated on diluted effluent neglecting domestic sewage flow) with final settlement tanks 20 000 gallons. Table 12.1 gives some results for 1960–64.

Surplus sludge is discharged to small, earth lagoons and removed by suction tanker. The main storage lagoon was cleaned out after five years continuous operation. The porous dome diffusers were removed once in five years, the aeration tanks emptied occasionally and the diffusers washed externally.

TABLE 12.2
Papermill activated-sludge plant average 1960–61

		Average results in parts per million			
Feed (*from main lagoon*)			Effluent		
SS	PV	BOD	SS	PV	BOD
63	80	170	21	29	24

A mechanical aeration type plant (Ames Crosta Simplex) has been in operation at Colthrop since about 1924, and it must have been one of the very early applications of the process to the treatment of trade wastes. The plant consists of four combined aeration and settlement units 30 ft square, 13 ft depth, with a total aeration volume of 154 000 gal and the corner settlement pockets 46 000 gal. Arrangements were made to feed the plant with waste drawn from the main effluent lagoon system. The plant can probably deal with 200 000 gal/day of waste with average BOD of 200 p.p.m. Table 12.2 gives average results for twelve months to May 1961.

In the summer months the average BOD can be below 5 p.p.m. but there is a deterioration in the winter period possibly due to the low temperature reached by the stored water in the large lagoon. This temperature effect has not been evident in the results obtained with the Bridge Mill plant but is usually experienced with the Colthrop percolating filters.

Two new activated-sludge plants (Ames Crosta) were built, costing nearly £80 000, and brought into operation in March 1963. The two plants are identical, each designed to treat 400 000 gal/day of effluent from the main lagoon with a BOD of 270 p.p.m. The waste from the lagoon is screened, to remove floating pieces of polythene, etc., by a Longwood Engineering rotary brush screen and pumped into an inlet channel between the two plants.

The aeration tanks (400 000 gal) 100 × 60 ft overall size, comprise eight, 25 ft square aeration pockets 12·5 ft deep with the high-intensity aeration cones driven in pairs through line shafts with chain drives allowing a choice of three speeds. The submergence of the cones can also be varied from 0–9 but is normally set at $4\frac{1}{2}$ in. The four settlement tanks which are 40 ft in diameter and 6 ft deep have a total capacity of 190 000 gal. Return activated sludge is pumped back to both inlet cones with surplus discharged to a series of three earth lagoons, the supernatant overflowing to the inlet of the main lagoon. Probably due to the septic nature of the waste from the lagoon and the very low suspended solids present, difficulty was experienced in building up a supply of activated sludge. A month was required before the BOD was reduced by 50% and a further three weeks to produce effluents with BOD's below 20 p.p.m. With the plant treating 800 000 gal of lagoon waste per day the effluent was of a satisfactory quality with the BOD below 10 p.p.m. at times and probably averaging 15 p.p.m. for the seven months from the beginning of May 1963. The performance of the plant deteriorated in December possibly due to overloading caused by an increase in strength of the waste and a faulty meter combined with excessive withdrawal of surplus sludge. Difficulty was again experienced in building up a new supply of activated sludge, and the plant did not fully recover until the spring. The low growth rate of solids has meant that the quantities of surplus sludge produced have been well below the quantities to be expected from operation of a normal sewage plant. The optimum activated-sludge concentration in the aeration tanks is probably between 2–3000 p.p.m. corresponding to one hour settlement figures of 50–60%. Foaming has only been a problem when the aeration tank solids concentration has been low. The effluent can be given further treatment by discharging it through a series of shallow, earth lagoons.

Variations in plant design using combined aeration settlement tanks have been used in USA and in Germany and Paterson Candy International have installed one of their 55-ft-diameter 'Biological Clarifiers' at Aylesford Papermills to treat waste similar in characteristics to the waste treated at Tovil and Bridge. This type of plant has been described by Punt and Cook.

BIBLIOGRAPHY

Waldemeyer, T., 'Purification of Papermill Effluents by Activated Sludge', *Proc. Tech. Sect. BP & BMA* (1958), **39** (3), 425–40. (These Proceedings also contain a series of other papers on 'Effluent Problems in the Paper Industry'.)

Waldemeyer, T., 'Disposal of Solids in the Paper Industry', *J. Inst. Sew. Purif.* (1962), **61**, 439.

Waldemeyer, T., 'Biological Treatment of Paper and Boardmill Effluents', *The Paper Maker*, October 1964, 96–102. (This supplement includes three other papers on 'Water and Effluent Treatment'.) Also November 1964: 'Use of synthetic plastic filter media for the biological treatment of papermill wastes'.

Punt, S. E., and Cook, W. J. M., 'Treatment and Recovery of Paper Mill Effluents', *Effluent and Water Treatment Manual*, 3rd ed. 1966, Thunderbird Enterprise Ltd, London.

Southgate, B. A., 'Treatment and Disposal of Waste Waters from Paper Mills', *Proc. Papermakers' Assoc. G.B.I.* (1945), **26**, 339.

INDEX

Toxic
  metals
    inhibition of biological pro-
      cesses by, 28, 72
    removal of by sewage, 29
    toxic concentration of, 31
    waters, in, 27, 33
  substances
    textile wastes, in, 192
    waters, in, 7, 8, 145
Toxicity, 9, 29
  addition of, 31
Treatability factor for farm
      effluents, 139

Units
  conversion factors, xiii
  inconstancy of, xi

Vacuum filters, 200
Vegetable
  peeling, 90
  washing, 88, 89

Waste disposal in Nature, 13
Water
  hardness, effect on
    ion-exchangers, 158
    toxicity, 8, 31
  naturally toxic, 9
  pollution, definition of, 1
  self-purification of, 3
  transport, 88
  usage in paper manufacture, 197,
      198
Wedge-wire beds, 101
Wilting, 131
Wood pulping, BOD loads from,
      197
Wool, 187
  scouring wastes, 190

Zinc
  inhibition of sludge digestion by,
      72
  mining, water pollution by, 28
  permissible concentration of in
      waters, 7, 8
  plating solutions, 167